student stud
ART NOTEBOOK

to accompany

Essentials of
Human Anatomy and Physiology
Sixth Edition

David Shier

◆

Jackie Butler

◆

Ricki Lewis

McGraw Hill **WCB McGraw-Hill**

Boston, Massachusetts   Burr Ridge, Illinois   Dubuque, Iowa
Madison, Wisconsin   New York, New York   San Francisco, California   St. Louis, Missouri

The McGraw-Hill Companies Higher Education Group
A Division of The McGraw-Hill Companies

Student Study Art Notebook
to accompany Shier et al, *Hole's Essentials of Human Anatomy and Physiology, 6e.*

This book is printed on recycled paper.

1 2 3 4 5 6 7 8 9 0 QPD QPD 9 0 9 8 7

ISBN 0-697-32919-4

TO THE STUDENT

The *Student Study Art Notebook* is designed to help you in your study of human anatomy and physiology. The notebook contains art reproduced from the textbook. Each figure also corresponds to one of the 250 overhead transparencies; thus you can take notes during lectures, or jot down comments as you are reading through the chapters.

The notebook is perforated and 3-hole punched, so if you wish, you can remove sheets and put them in a binder with other study or lecture notes. Any blank pages at the end of this notebook can be used for additional notes or drawings.

We hope this notebook, used along with your text, helps to make the study of the human body easier for you.

DIRECTORY OF NOTEBOOK FIGURES

TO ACCOMPANY

SHIER/BUTLER/LEWIS

HOLE'S ESSENTIALS OF HUMAN ANATOMY AND PHYSIOLOGY, 6E.

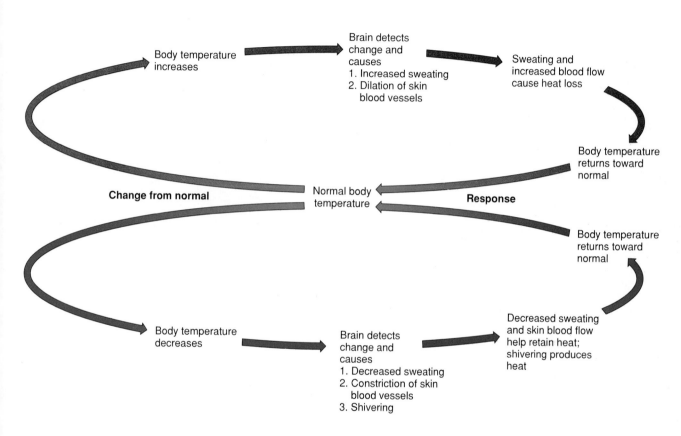

Homeostatic Mechanism
Figure 1.4

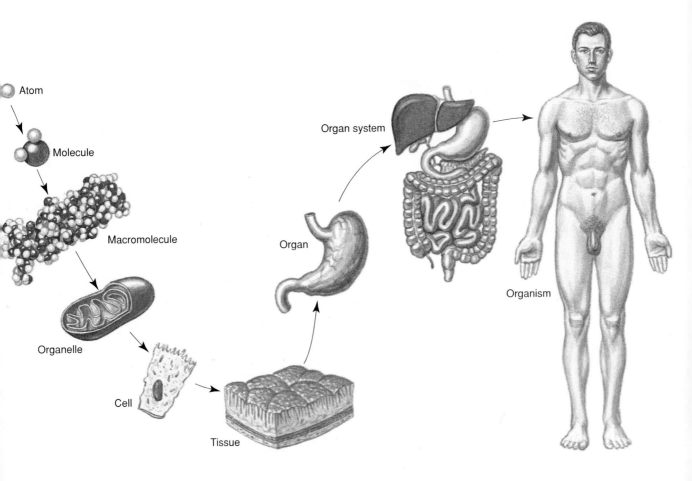

Atom

Molecule

Macromolecule

Organelle

Cell

Tissue

Organ

Organ system

Organism

Hierarchy of Organization
Figure 1.5

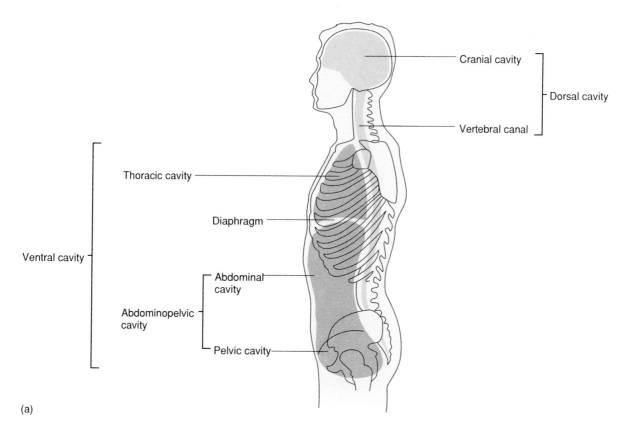

(a)

Body Cavities (lateral)
Figure 1.6a

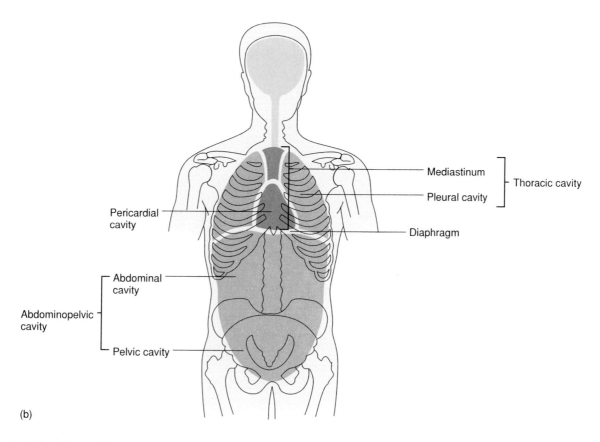

(b)

Body Cavities (frontal)
Figure 1.6b

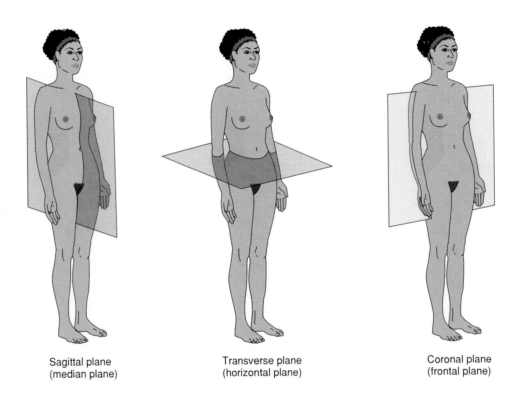

Sagittal plane
(median plane)

Transverse plane
(horizontal plane)

Coronal plane
(frontal plane)

Body Planes
Figure 1.10

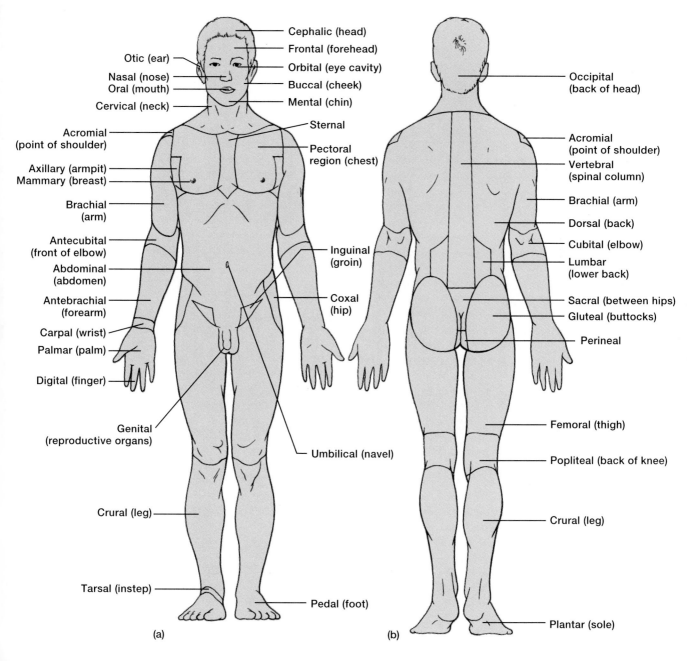

Cephalic (head)
Frontal (forehead)
Otic (ear)
Orbital (eye cavity)
Nasal (nose)
Oral (mouth)
Buccal (cheek)
Cervical (neck)
Mental (chin)
Acromial
(point of shoulder)
Sternal
Pectoral
region (chest)
Axillary (armpit)
Mammary (breast)
Brachial
(arm)
Antecubital
(front of elbow)
Inguinal
(groin)
Abdominal
(abdomen)
Antebrachial
(forearm)
Coxal
(hip)
Carpal (wrist)
Palmar (palm)
Digital (finger)
Genital
(reproductive organs)
Umbilical (navel)
Crural (leg)
Tarsal (instep)
Pedal (foot)

Occipital
(back of head)
Acromial
(point of shoulder)
Vertebral
(spinal column)
Brachial (arm)
Dorsal (back)
Cubital (elbow)
Lumbar
(lower back)
Sacral (between hips)
Gluteal (buttocks)
Perineal
Femoral (thigh)
Popliteal (back of knee)
Crural (leg)
Plantar (sole)

(a)

(b)

Body Regions
Figure 1.13

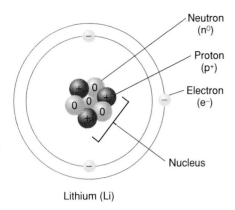

Neutron
(n⁰)

Proton
(p⁺)

Electron
(e⁻)

Nucleus

Lithium (Li)

Organization of an Atom
Figure 2.1

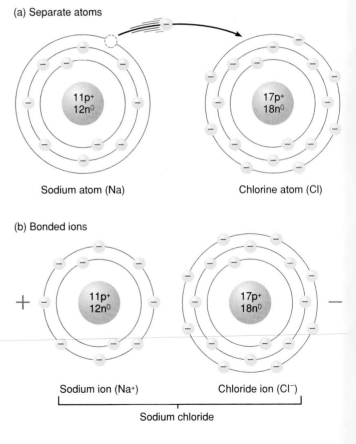

(a) Separate atoms

Sodium atom (Na)   Chlorine atom (Cl)

(b) Bonded ions

Sodium ion (Na⁺)   Chloride ion (Cl⁻)

Sodium chloride

Ionic Bond
Figure 2.4

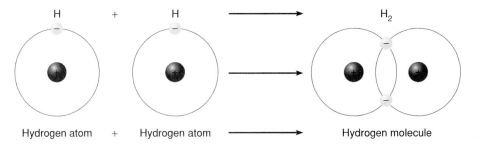

Covalent Bond
Figure 2.5

H
‖
C
|
H — C — O — H
|
H — O — C — H
|
H — C — O — H
|
H — C — O — H
|
H — C — O — H
|
H

(a)

H
|
H — C — O — H
|
C   O
H   H   H
| | |
C   O — H   C
H — O | | O — H
        C   C
        | |
        H   O — H

(b)

O

(c)

pH Values of Common Substances
Figure 2.10

(a) Monosaccharide   (b) Disaccharide

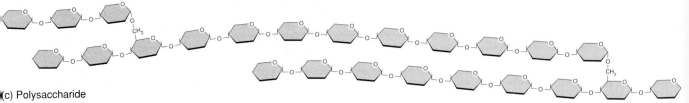

(c) Polysaccharide

Structure of Glucose
Figure 2.11

Amino acid	Structural formula
Alanine	Amino group $\quad$ Carboxyl group $\quad$ H−N−C−C with H, H on nitrogen and carbon, =O and OH on carboxyl, H−C−H, H below
Valine	H−N−C−C=O, OH; CH; H−C−H and H−C−H; H, H
Cysteine	H−N−C−C=O, OH; H−C−H; SH

Structure of a Triglyceride
Figure 2.13

8

Structure of an Amino Acid
Figure 2.14

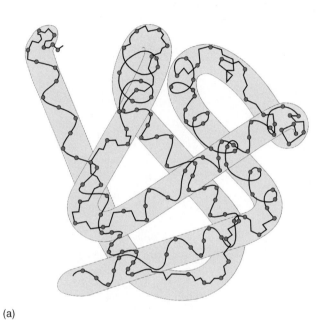

(a)

Secondary Structure of a Protein
Figure 2.16a

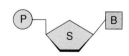

Tertiary Structure of a Protein
Figure 2.17

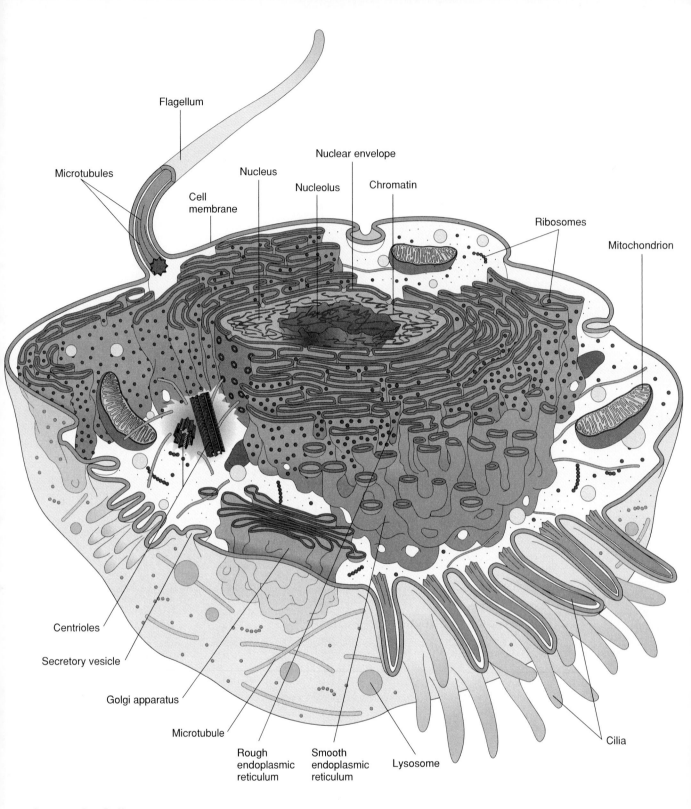

Flagellum

Microtubules

Cell
membrane

Nucleus

Nuclear envelope

Nucleolus

Chromatin

Ribosomes

Mitochondrion

Centrioles

Secretory vesicle

Golgi apparatus

Microtubule

Rough
endoplasmic
reticulum

Smooth
endoplasmic
reticulum

Lysosome

Cilia

Composite Cell
Figure 3.2

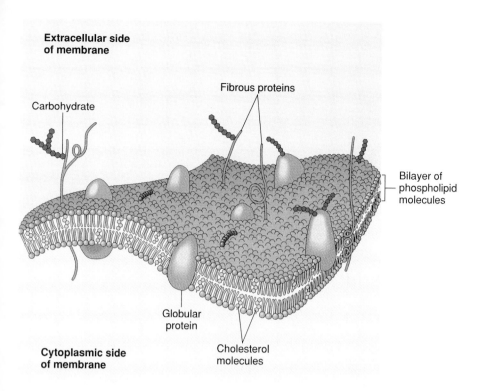

Extracellular side
of membrane

Carbohydrate

Fibrous proteins

Bilayer of
phospholipid
molecules

Globular
protein

Cholesterol
molecules

Cytoplasmic side
of membrane

Cell Membrane
Figure 3.3

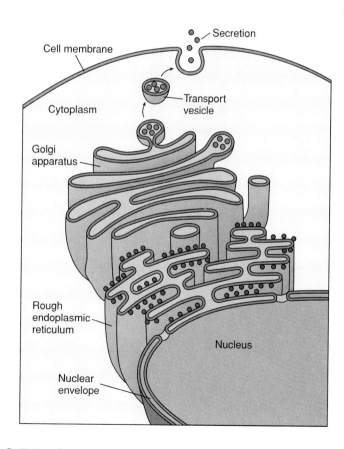

Secretion

Cell membrane

Cytoplasm

Transport
vesicle

Golgi
apparatus

Rough
endoplasmic
reticulum

Nucleus

Nuclear
envelope

Cellular Secretion
Figure 3.5

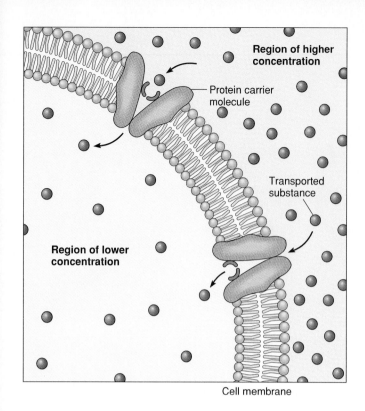

Region of higher concentration

Protein carrier molecule

Transported substance

Region of lower concentration

Cell membrane

Facilitated Diffusion
Figure 3.13

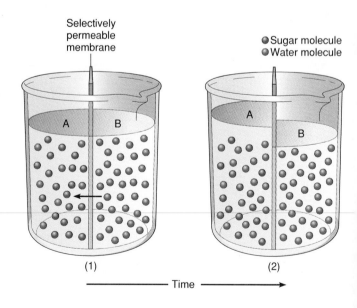

Selectively permeable membrane

● Sugar molecule
● Water molecule

A B

A B

(1)

(2)

Time

Osmosis
Figure 3.14

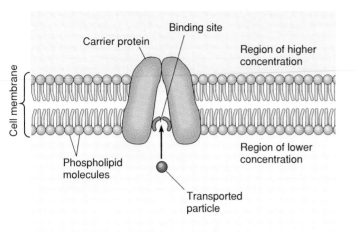

(a)

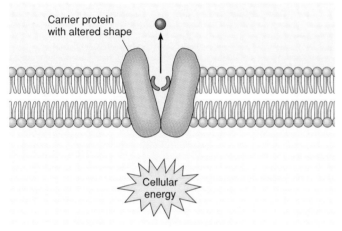

(b)

Active Transport
Figure 3.16

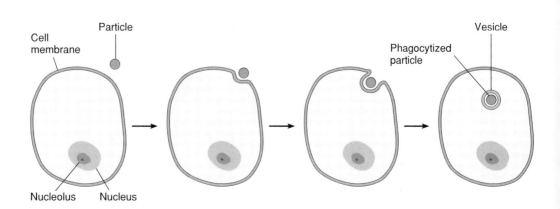

Phagocytosis
Figure 3.17

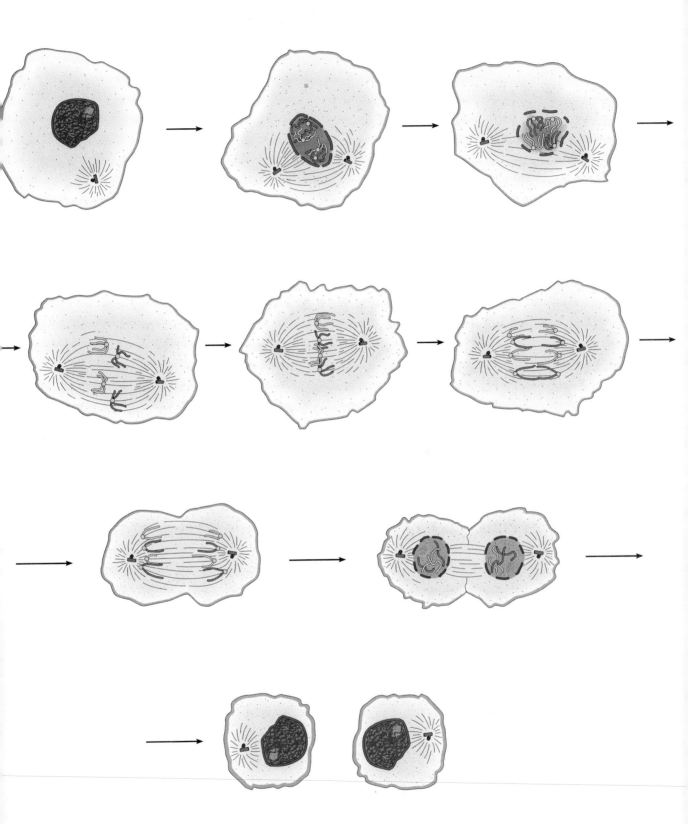

Mitosis
Figure 3.18

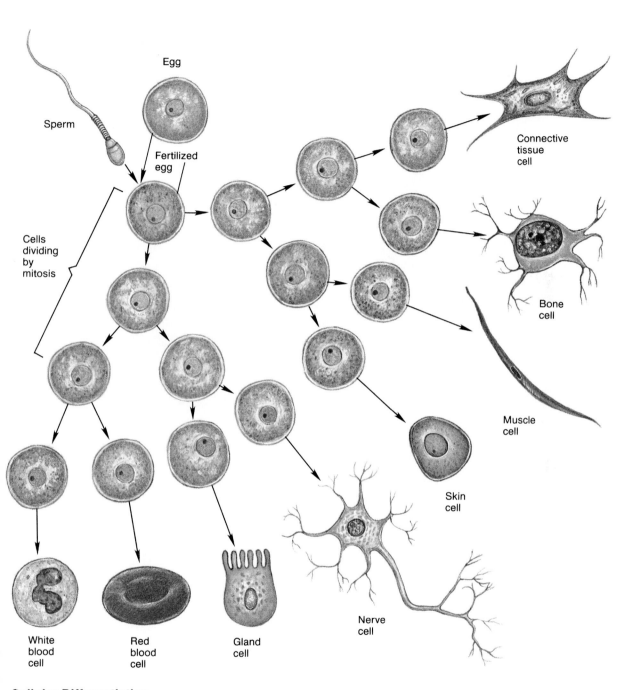

Egg

Sperm

Fertilized egg

Cells dividing by mitosis

Connective tissue cell

Bone cell

Muscle cell

Skin cell

Nerve cell

White blood cell

Red blood cell

Gland cell

Cellular Differentiation
Figure 3.20

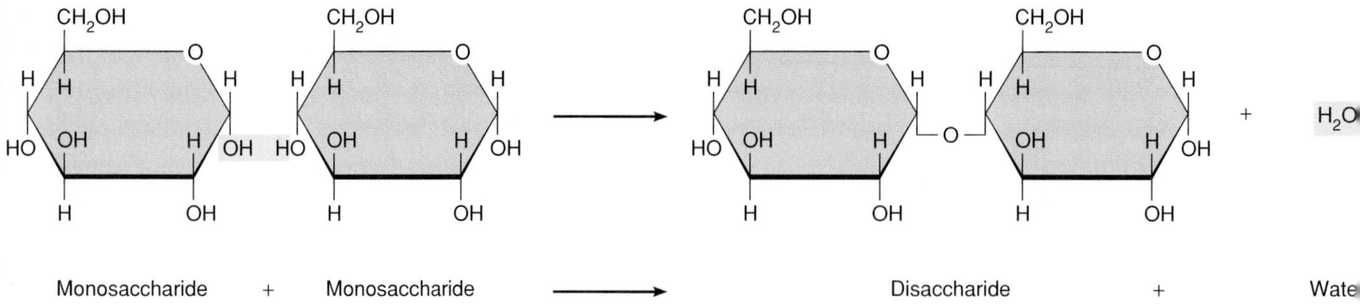

Dehydration Synthesis Forms Disaccharide
Figure 4.1

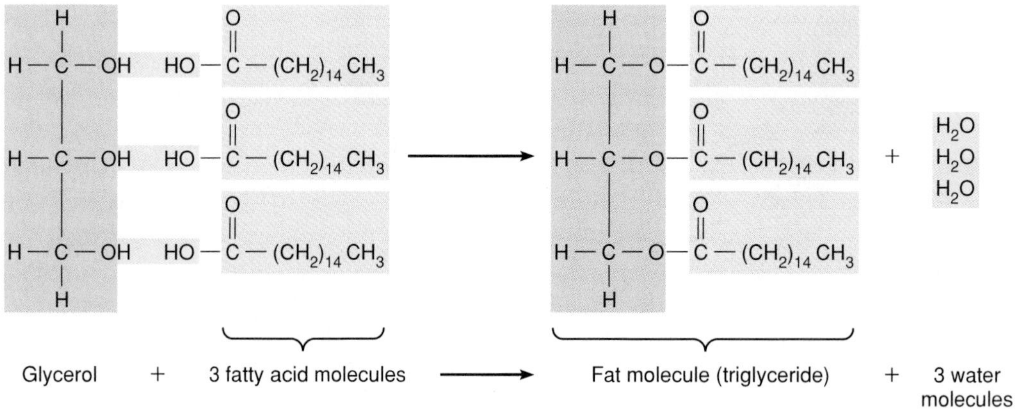

Dehydration Synthesis Forms Fat
Figure 4.2

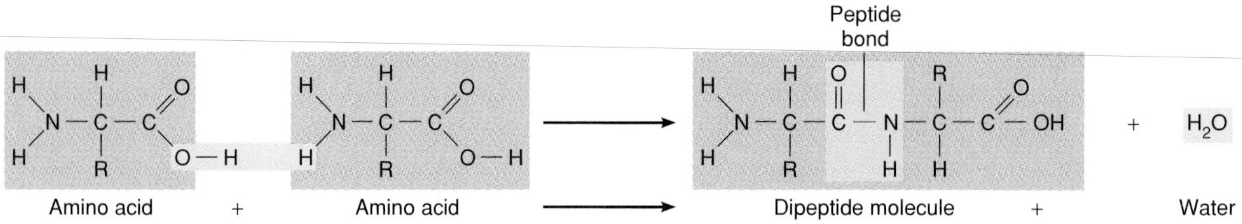

Dehydration Synthesis Forms Dipeptide
Figure 4.3

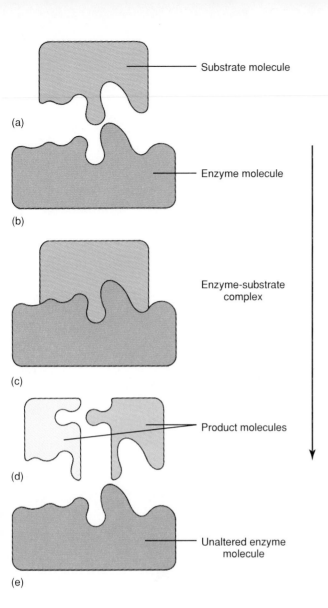

Substrate molecule

(a)

Enzyme molecule

(b)

Enzyme-substrate
complex

(c)

Product molecules

(d)

Unaltered enzyme
molecule

(e)

Enzyme-Substrate Interaction
Figure 4.4

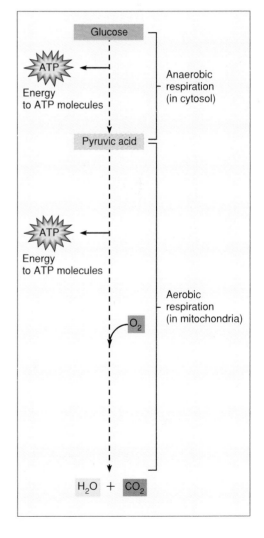

Glucose

ATP

Energy
to ATP molecules

Anaerobic
respiration
(in cytosol)

Pyruvic acid

ATP

Energy
to ATP molecules

Aerobic
respiration
(in mitochondria)

O_2

H_2O + CO_2

Overview of Cellular Metabolism
Figure 4.5

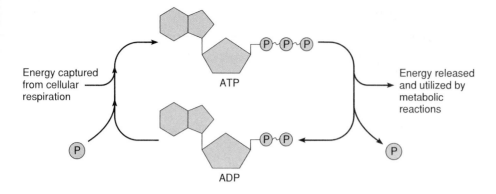

ATP
Figure 4.6

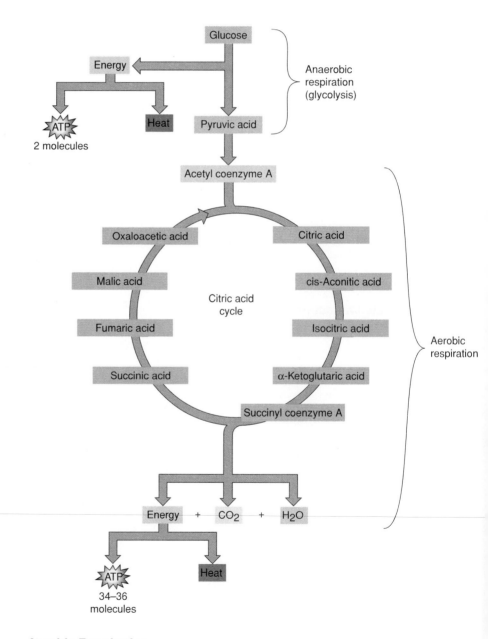

Aerobic Respiration
Figure 4.8

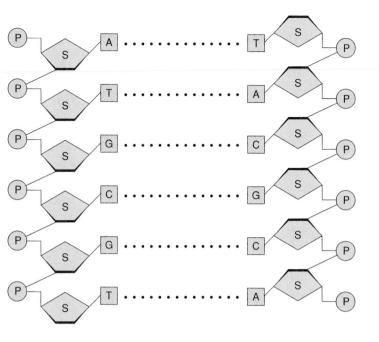

Base-Pairing of DNA Bases
Figure 4.12

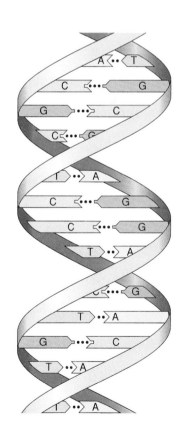

Portion of a DNA Double Helix
Figure 4.13

DNA   RNA

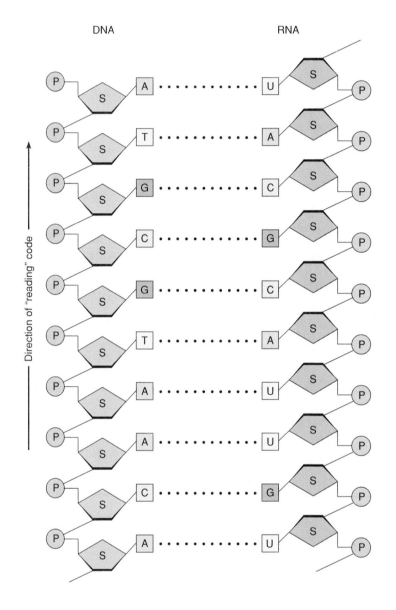

Base-Pairing of DNA and RNA Bases
Figure 4.15

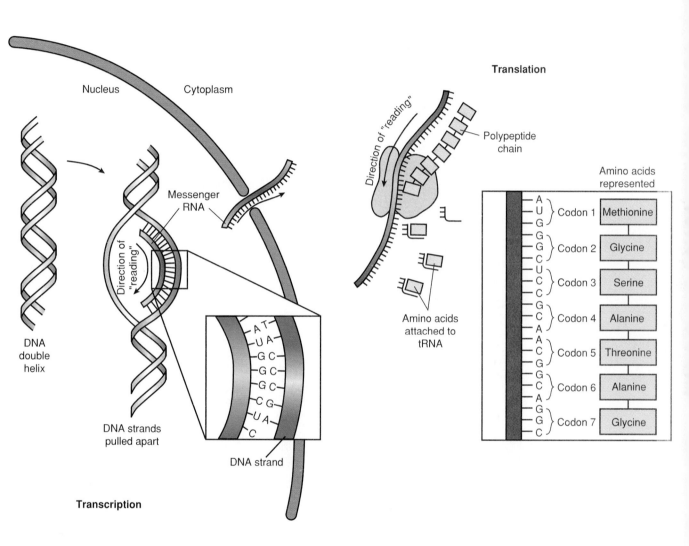

Nucleus   Cytoplasm

Translation

Direction of "reading"

Messenger
RNA

Polypeptide
chain

Direction of
"reading"

Amino acids
represented

DNA
double
helix

Amino acids
attached to
tRNA

A
U   } Codon 1   Methionine
G

G
G   } Codon 2   Glycine
C

U
C   } Codon 3   Serine
C

G
C   } Codon 4   Alanine
A

A
C   } Codon 5   Threonine
G

G
C   } Codon 6   Alanine
A

G
G   } Codon 7   Glycine
C

DNA strands
pulled apart

A T
U A
G C
G C
G C
C G
U A
C

DNA strand

Transcription

Transcription and Translation
Figure 4.17

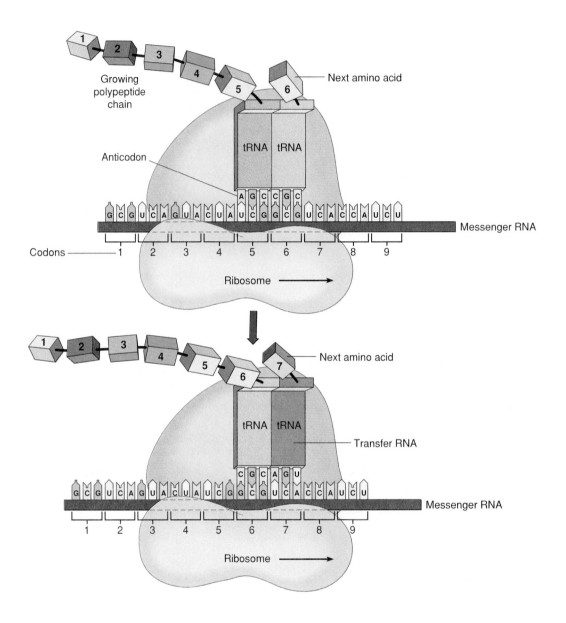

Protein Synthesis
Figure 4.18

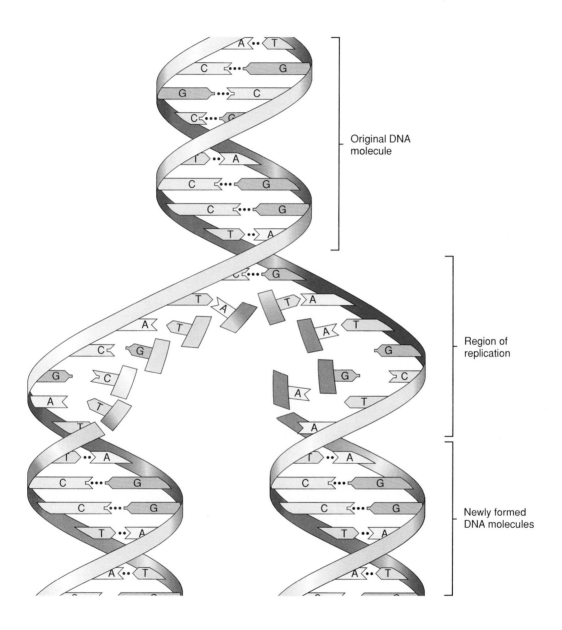

DNA Replication
Figure 4.19

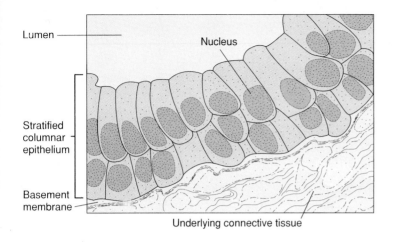

Lumen

Nucleus

Stratified
columnar
epithelium

Basement
membrane

Underlying connective tissue

Types of Exocrine Glands
Figure 5.7

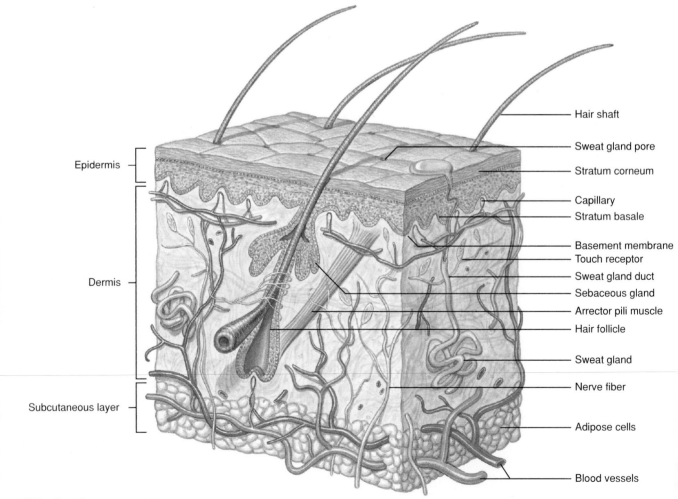

Epidermis

Dermis

Subcutaneous layer

Hair shaft

Sweat gland pore

Stratum corneum

Capillary

Stratum basale

Basement membrane

Touch receptor

Sweat gland duct

Sebaceous gland

Arrector pili muscle

Hair follicle

Sweat gland

Nerve fiber

Adipose cells

Blood vessels

Skin Section
Figure 6.1

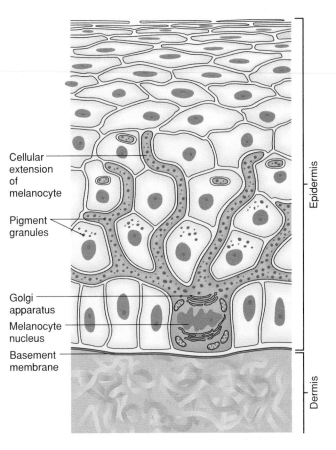

Cellular
extension
of
melanocyte

Pigment
granules

Golgi
apparatus

Melanocyte
nucleus

Basement
membrane

Epidermis

Dermis

A Melanocyte
Figure 6.3

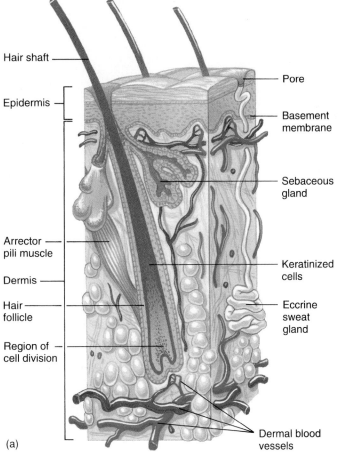

Hair shaft

Epidermis

Arrector
pili muscle

Dermis

Hair
follicle

Region of
cell division

Pore

Basement
membrane

Sebaceous
gland

Keratinized
cells

Eccrine
sweat
gland

Dermal blood
vessels

(a)

Hair Follicle
Figure 6.4a

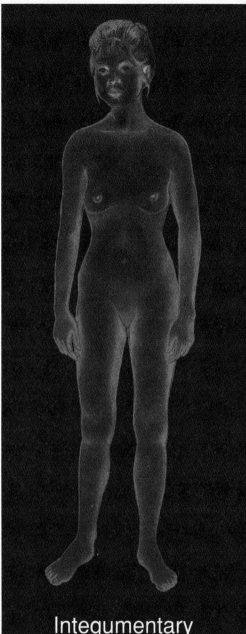

Integumentary System

The skin provides protection, contains sensory organs, and helps control body temperature.

Skeletal System

Vitamin D activated by the skin helps provide calcium for bone matrix.

Muscular System

Involuntary muscle contractions (shivering) work with the skin to control body temperature. Muscles act on facial skin to create expressions.

Nervous System

Sensory receptors provide information about the outside world to the nervous system. Nerves control the activity of sweat glands.

Endocrine System

Hormones help to increase skin blood flow during exercise. Other hormones stimulate either the synthesis or the decomposition of subcutaneous fat.

Cardiovascular System

Skin blood vessels play a role in regulating body temperature.

Lymphatic System

The skin provides an important first line of defense for the immune system.

Digestive System

Excess calories may be stored as subcutaneous fat. Vitamin D activated by the skin stimulates dietary calcium absorption.

Respiratory System

Stimulation of skin receptors may alter respiratory rate.

Urinary System

The kidneys help compensate for water and electrolytes lost in sweat.

Reproductive System

Sensory receptors play an important role in sexual activity and in the suckling reflex.

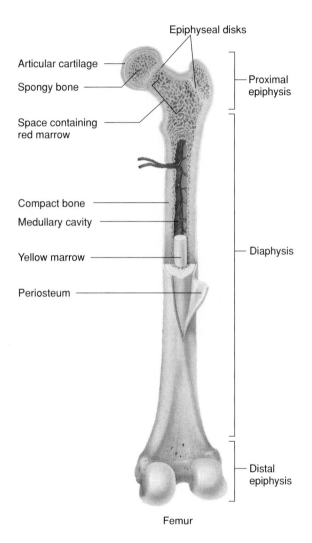

Epiphyseal disks

Articular cartilage

Spongy bone

Space containing
red marrow

Compact bone

Medullary cavity

Yellow marrow

Periosteum

Proximal
epiphysis

Diaphysis

Distal
epiphysis

Femur

Structure of a Long Bone
Figure 7.1

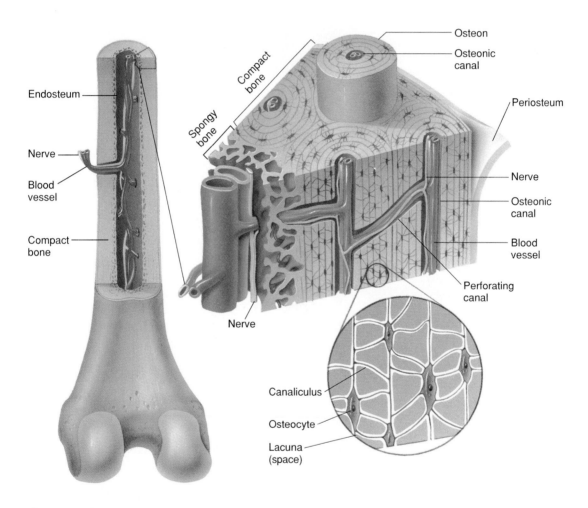

Endosteum

Nerve

Blood
vessel

Compact
bone

Spongy
bone

Compact
bone

Osteon

Osteonic
canal

Periosteum

Nerve

Osteonic
canal

Blood
vessel

Perforating
canal

Nerve

Canaliculus

Osteocyte

Lacuna
(space)

Compact Bone
Figure 7.3

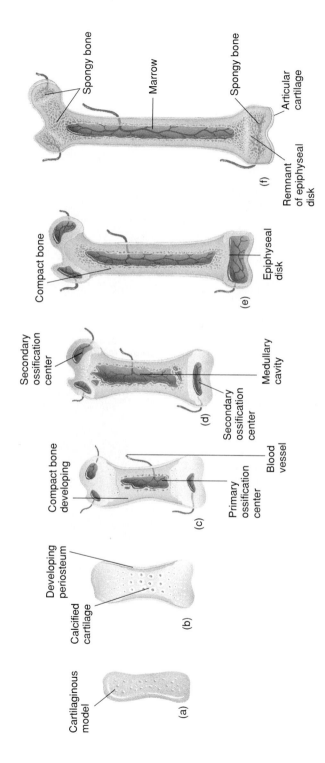

Development of an Endochondral Bone
Figure 7.5

29

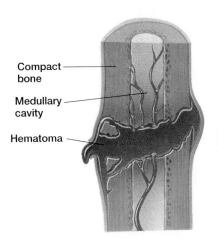

Compact
bone

Medullary
cavity

Hematoma

(a) Blood escapes from ruptured
blood vessels and forms a
hematoma.

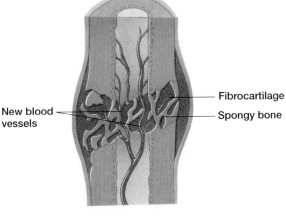

New blood
vessels

Fibrocartilage

Spongy bone

(b) Spongy bone forms in regions
close to developing blood vessels,
and fibrocartilage forms in more
distant regions.

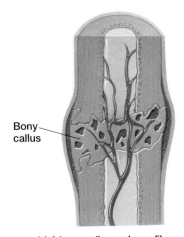

Bony
callus

(c) A bony callus replaces fibrocartilage.

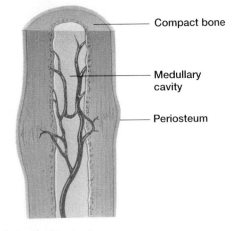

Compact bone

Medullary
cavity

Periosteum

(d) Osteoclasts remove excess
bony tissue, restoring new bone
structure much like the original.

Fracture Repair
Box 7A

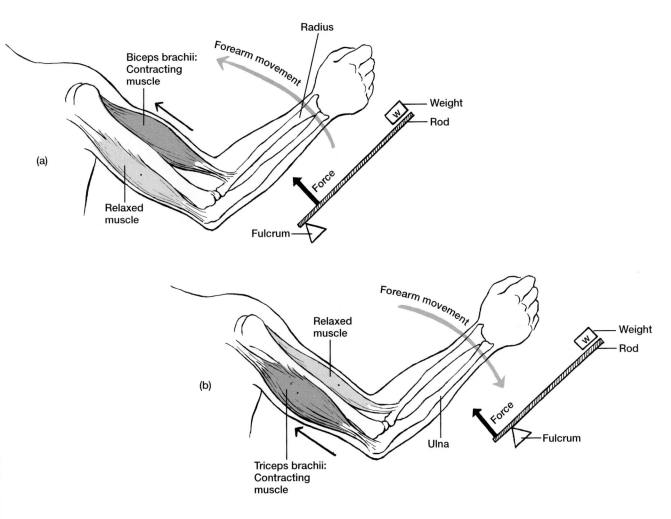

Radius

Biceps brachii:
Contracting
muscle

Forearm movement

Weight
Rod

(a)

Force

Relaxed
muscle

Fulcrum

Forearm movement

Relaxed
muscle

Weight
Rod

(b)

Force

Fulcrum

Ulna

Triceps brachii:
Contracting
muscle

Limb and Lever
Figure 7.6

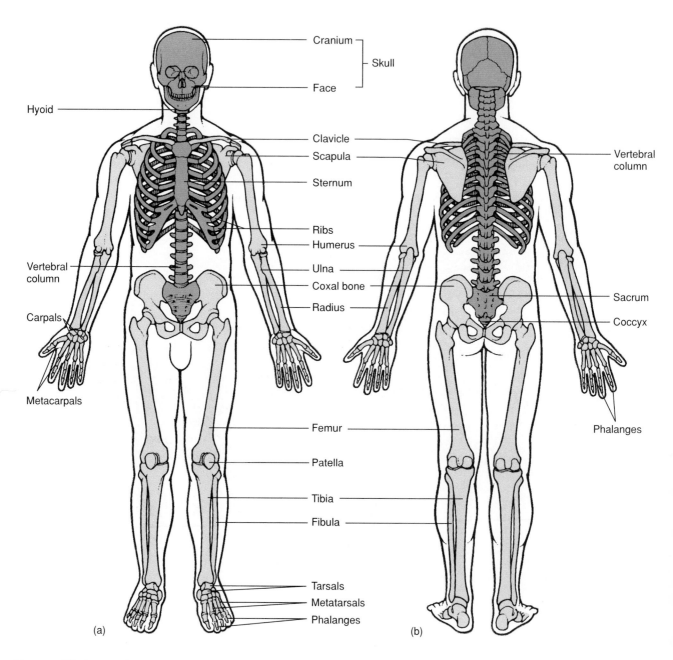

Cranium — Skull
Face

Hyoid

Clavicle
Scapula
Sternum

Ribs
Humerus

Vertebral column

Ulna
Coxal bone
Radius

Carpals

Metacarpals

Femur

Patella

Tibia

Fibula

Tarsals
Metatarsals
Phalanges

(a)

Vertebral column

Sacrum

Coccyx

Phalanges

(b)

Human Skeleton, Anterior and Posterior
Figure 7.7

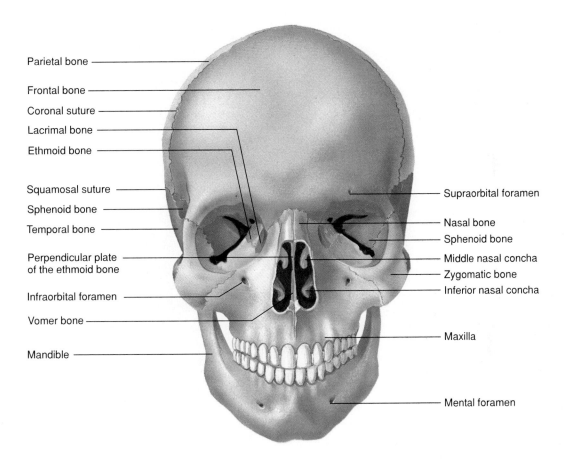

Parietal bone

Frontal bone

Coronal suture

Lacrimal bone

Ethmoid bone

Squamosal suture

Sphenoid bone

Temporal bone

Perpendicular plate
of the ethmoid bone

Infraorbital foramen

Vomer bone

Mandible

Supraorbital foramen

Nasal bone

Sphenoid bone

Middle nasal concha

Zygomatic bone

Inferior nasal concha

Maxilla

Mental foramen

Human Skull, Anterior View
Figure 7.8

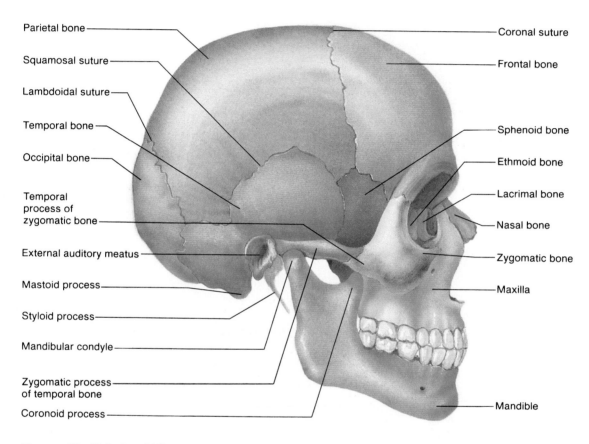

Parietal bone —————————

Squamosal suture —————

Lambdoidal suture —————

Temporal bone —————

Occipital bone —————

Temporal
process of
zygomatic bone —————

External auditory meatus —————

Mastoid process —————

Styloid process —————

Mandibular condyle —————

Zygomatic process —————
of temporal bone

Coronoid process —————

————— Coronal suture

————— Frontal bone

————— Sphenoid bone

————— Ethmoid bone

————— Lacrimal bone

————— Nasal bone

————— Zygomatic bone

————— Maxilla

————— Mandible

Human Skull, Lateral View
Figure 7.10

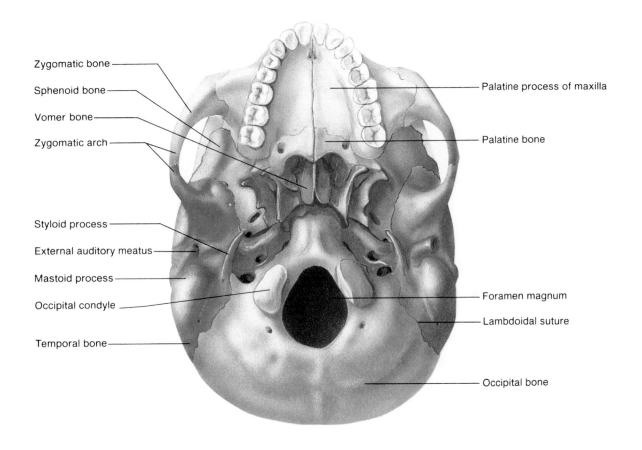

Zygomatic bone

Sphenoid bone

Vomer bone

Zygomatic arch

Styloid process

External auditory meatus

Mastoid process

Occipital condyle

Temporal bone

Palatine process of maxilla

Palatine bone

Foramen magnum

Lambdoidal suture

Occipital bone

Human Skull, Inferior View
Figure 7.11

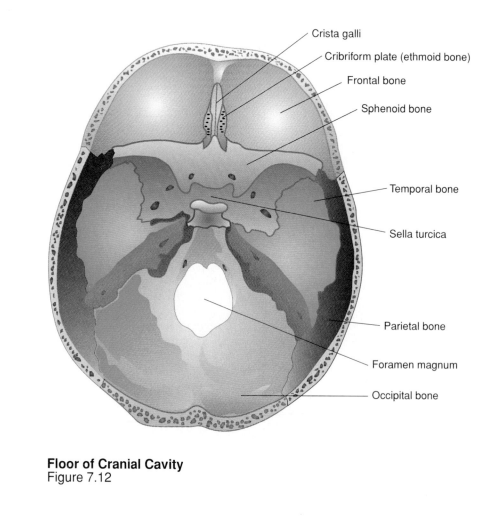

Crista galli

Cribriform plate (ethmoid bone)

Frontal bone

Sphenoid bone

Temporal bone

Sella turcica

Parietal bone

Foramen magnum

Occipital bone

Floor of Cranial Cavity
Figure 7.12

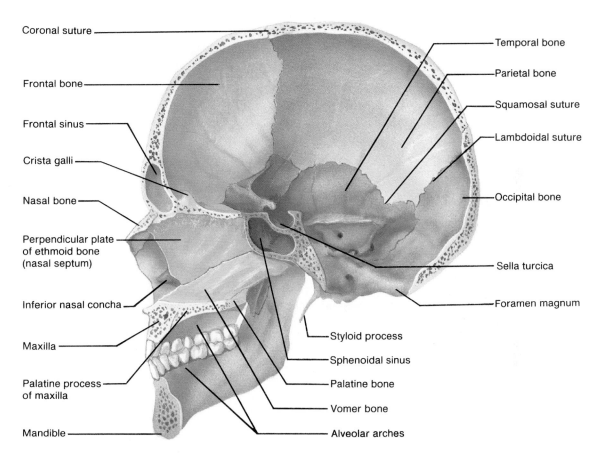

Coronal suture

Frontal bone

Frontal sinus

Crista galli

Nasal bone

Perpendicular plate
of ethmoid bone
(nasal septum)

Inferior nasal concha

Maxilla

Palatine process
of maxilla

Mandible

Temporal bone

Parietal bone

Squamosal suture

Lambdoidal suture

Occipital bone

Sella turcica

Foramen magnum

Styloid process

Sphenoidal sinus

Palatine bone

Vomer bone

Alveolar arches

Human Skull, Sagittal Section
Figure 7.13

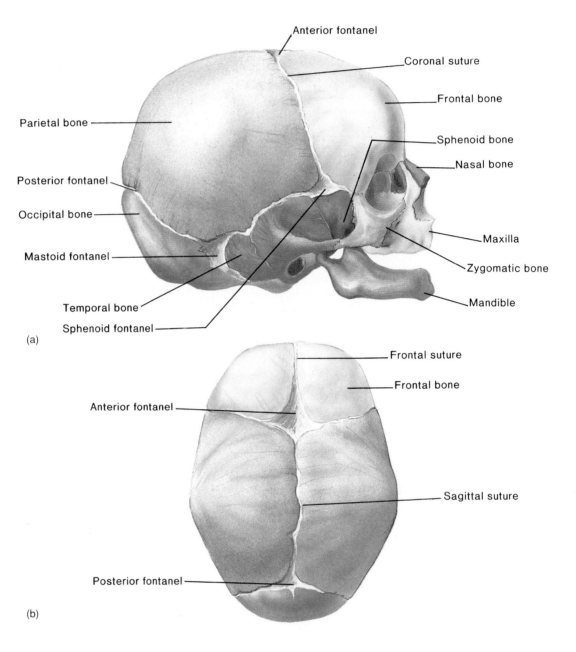

Anterior fontanel

Coronal suture

Frontal bone

Sphenoid bone

Nasal bone

Parietal bone

Posterior fontanel

Occipital bone

Maxilla

Mastoid fontanel

Zygomatic bone

Mandible

Temporal bone

Sphenoid fontanel

(a)

Frontal suture

Frontal bone

Anterior fontanel

Sagittal suture

Posterior fontanel

(b)

Infantile Skull
Figure 7.14

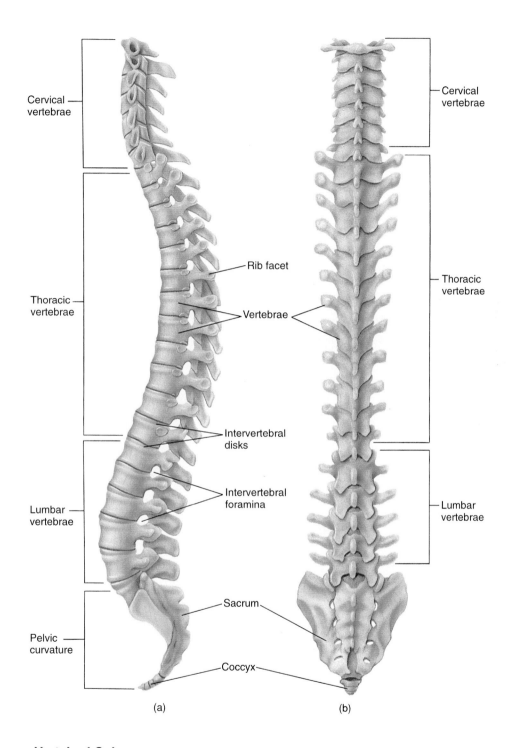

Cervical
vertebrae

Thoracic
vertebrae

Lumbar
vertebrae

Pelvic
curvature

Rib facet

Vertebrae

Intervertebral
disks

Intervertebral
foramina

Sacrum

Coccyx

Cervical
vertebrae

Thoracic
vertebrae

Lumbar
vertebrae

(a)

(b)

Vertebral Column
Figure 7.15

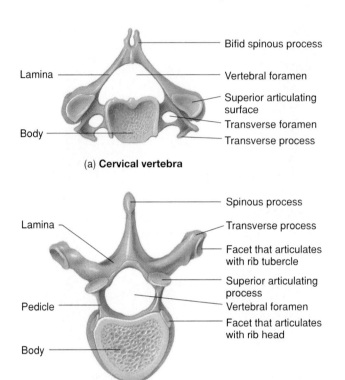

Lamina

Body

Bifid spinous process

Vertebral foramen

Superior articulating surface

Transverse foramen

Transverse process

(a) **Cervical vertebra**

Lamina

Pedicle

Body

Spinous process

Transverse process

Facet that articulates with rib tubercle

Superior articulating process

Vertebral foramen

Facet that articulates with rib head

(b) **Thoracic vertebra**

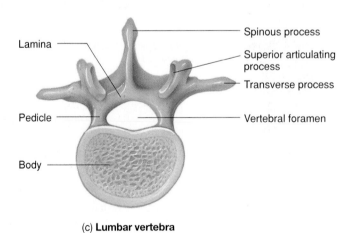

Lamina

Pedicle

Body

Spinous process

Superior articulating process

Transverse process

Vertebral foramen

(c) **Lumbar vertebra**

Types of Vertebrae
Figure 7.16

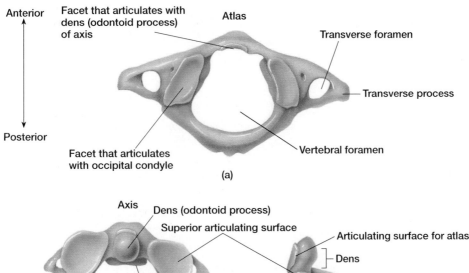

Anterior

Posterior

Facet that articulates with dens (odontoid process) of axis

Atlas

Transverse foramen

Transverse process

Facet that articulates with occipital condyle

Vertebral foramen

(a)

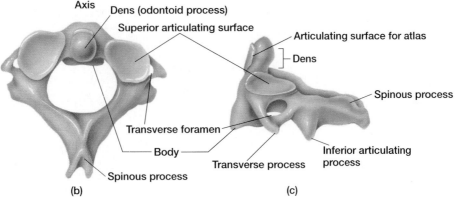

Axis

Dens (odontoid process)

Superior articulating surface

Articulating surface for atlas

Dens

Spinous process

Transverse foramen

Body

Spinous process

Transverse process

Inferior articulating process

(b)

(c)

Atlas and Axis
Figure 7.17

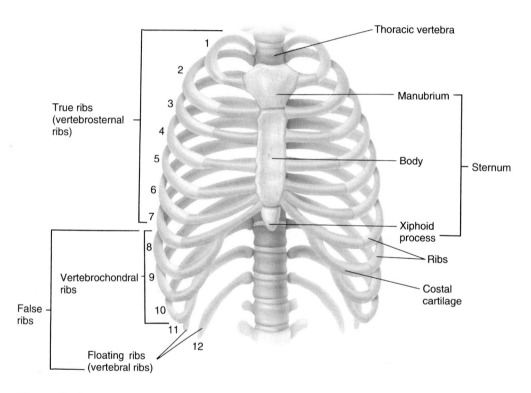

True ribs
(vertebrosternal
ribs)

1
2
3
4
5
6
7

Thoracic vertebra

Manubrium

Body

Sternum

Xiphoid
process

Ribs

Costal
cartilage

False
ribs

Vertebrochondral
ribs

8
9
10
11
12

Floating ribs
(vertebral ribs)

Thoracic Cage
Figure 7.19

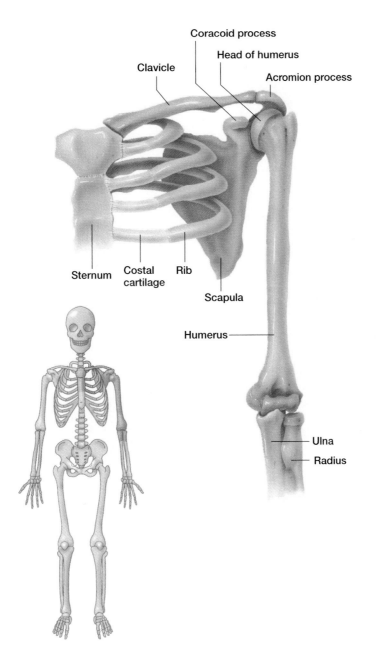

Coracoid process

Head of humerus

Clavicle

Acromion process

Sternum

Costal cartilage

Rib

Scapula

Humerus

Ulna

Radius

Pectoral Girdle
Figure 7.20

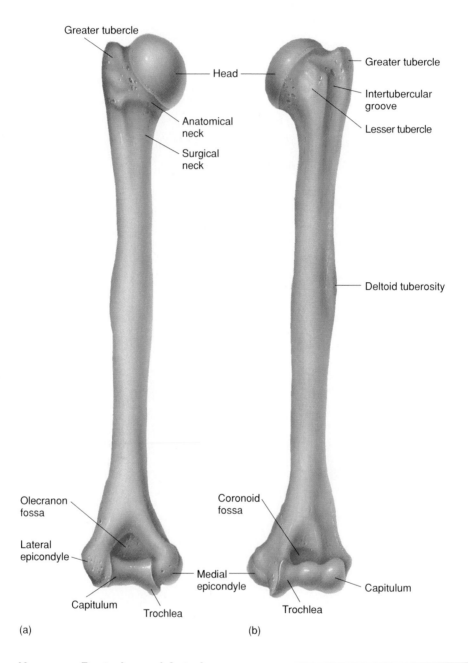

Greater tubercle

Head

Anatomical neck

Surgical neck

Greater tubercle

Intertubercular groove

Lesser tubercle

Deltoid tuberosity

Olecranon fossa

Coronoid fossa

Lateral epicondyle

Medial epicondyle

Capitulum

Capitulum

Trochlea

Trochlea

(a)

(b)

Humerus, Posterior and Anterior
Figure 7.23

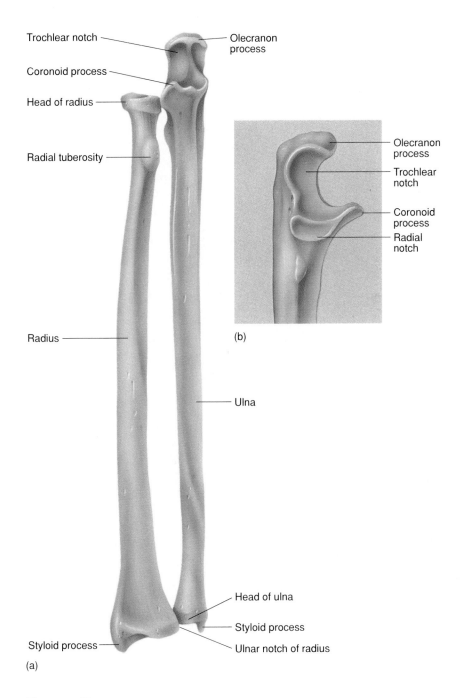

Trochlear notch

Coronoid process

Head of radius

Radial tuberosity

Radius

Olecranon process

Ulna

Olecranon process

Trochlear notch

Coronoid process

Radial notch

(b)

Head of ulna

Styloid process

Ulnar notch of radius

Styloid process

(a)

Forearm Bones
Figure 7.24

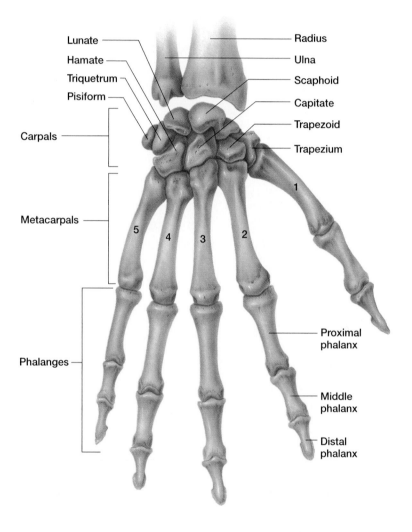

Lunate

Hamate

Triquetrum

Pisiform

Carpals

Metacarpals

Phalanges

Radius

Ulna

Scaphoid

Capitate

Trapezoid

Trapezium

1

5   4   3   2

Proximal
phalanx

Middle
phalanx

Distal
phalanx

Hand
Figure 7.25

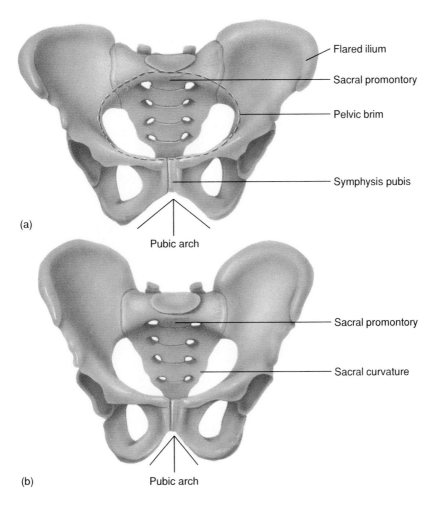

(a)

Flared ilium

Sacral promontory

Pelvic brim

Symphysis pubis

Pubic arch

(b)

Sacral promontory

Sacral curvature

Pubic arch

Pelvic Girdle, Female and Male
Figure 7.26

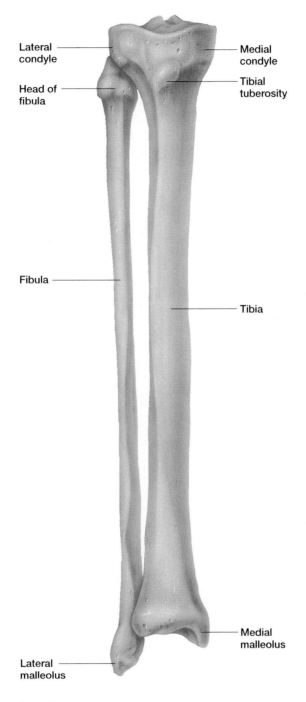

Lateral condyle

Head of fibula

Medial condyle

Tibial tuberosity

Fibula

Tibia

Medial malleolus

Lateral malleolus

Leg Bones
Figure 7.30

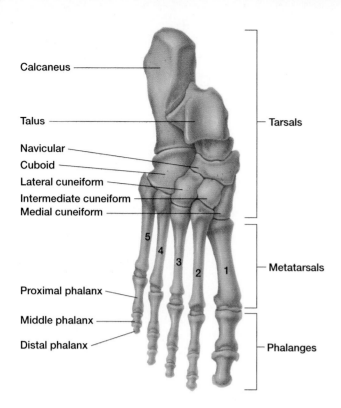

Calcaneus

Talus

Navicular

Cuboid

Lateral cuneiform

Intermediate cuneiform

Medial cuneiform

Tarsals

5

4

3

2

1

Metatarsals

Proximal phalanx

Middle phalanx

Distal phalanx

Phalanges

Foot
Figure 7.32

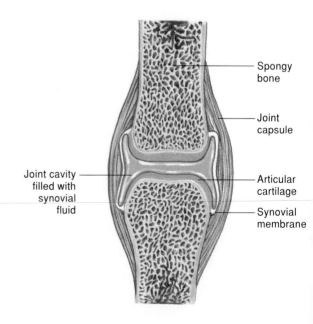

Spongy
bone

Joint
capsule

Joint cavity
filled with
synovial
fluid

Articular
cartilage

Synovial
membrane

Synovial Joint
Figure 7.34

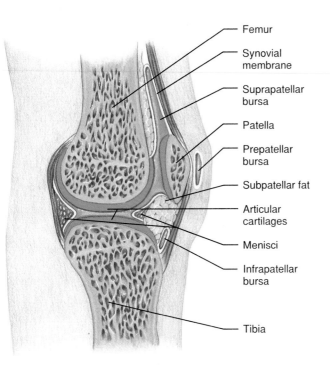

Femur

Synovial
membrane

Suprapatellar
bursa

Patella

Prepatellar
bursa

Subpatellar fat

Articular
cartilages

Menisci

Infrapatellar
bursa

Tibia

Knee Joint
Figure 7.35

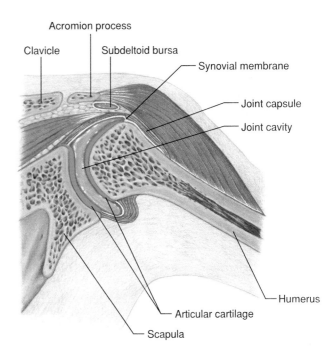

Acromion process

Clavicle

Subdeltoid bursa

Synovial membrane

Joint capsule

Joint cavity

Humerus

Articular cartilage

Scapula

Shoulder Joint
Figure 7.36

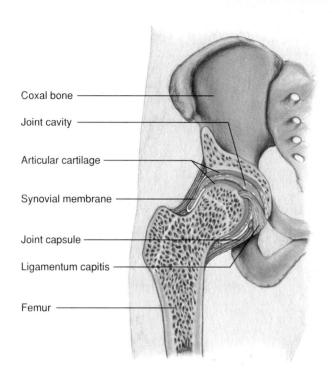

Coxal bone

Joint cavity

Articular cartilage

Synovial membrane

Joint capsule

Ligamentum capitis

Femur

Hip Joint
Figure 7.37

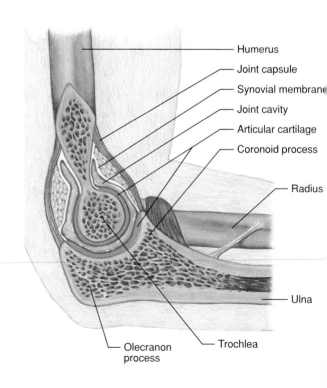

Humerus

Joint capsule

Synovial membrane

Joint cavity

Articular cartilage

Coronoid process

Radius

Ulna

Olecranon
process

Trochlea

Elbow Joint
Figure 7.38

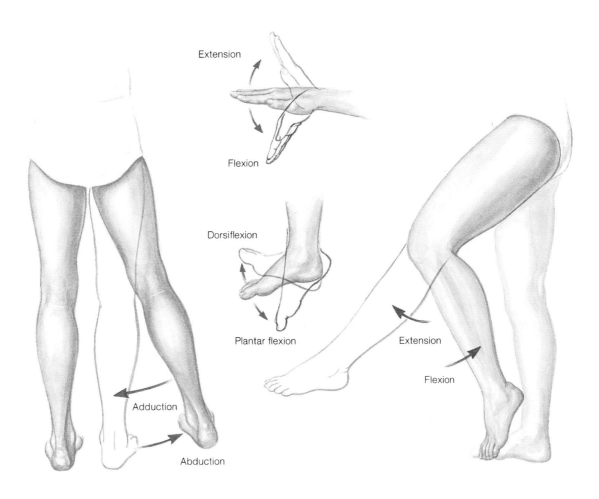

Extension

Flexion

Dorsiflexion

Plantar flexion

Extension

Flexion

Adduction

Abduction

Joint Movements I
Figure 7.39

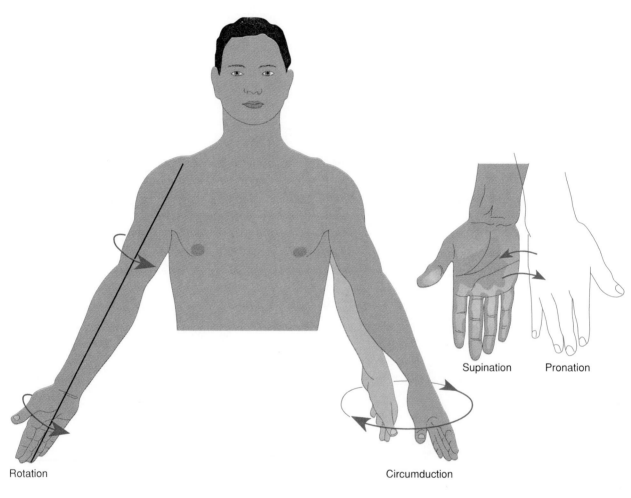

Rotation

Circumduction

Supination   Pronation

Joint Movements II
Figure 7.40

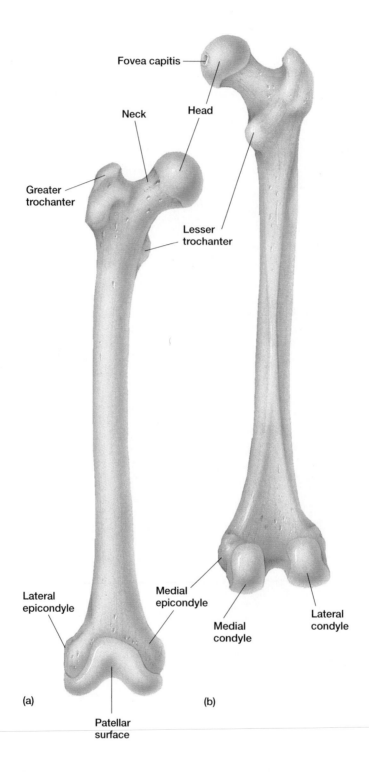

Fovea capitis

Neck   Head

Greater
trochanter

Lesser
trochanter

Lateral
epicondyle

Medial
epicondyle

Medial
condyle

Lateral
condyle

(a)   (b)

Patellar
surface

Thigh
Figure 7.29

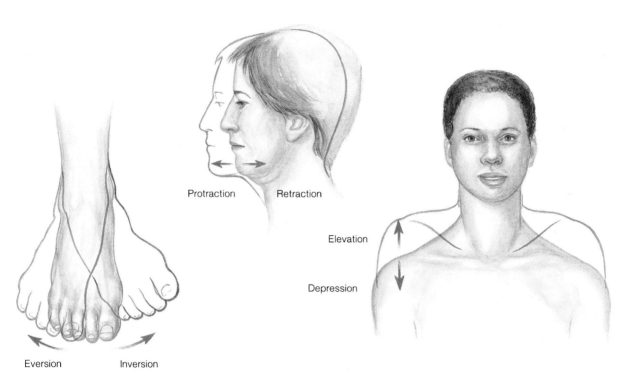

Eversion   Inversion

Protraction   Retraction

Elevation

Depression

Joint Movements III
Figure 7.41

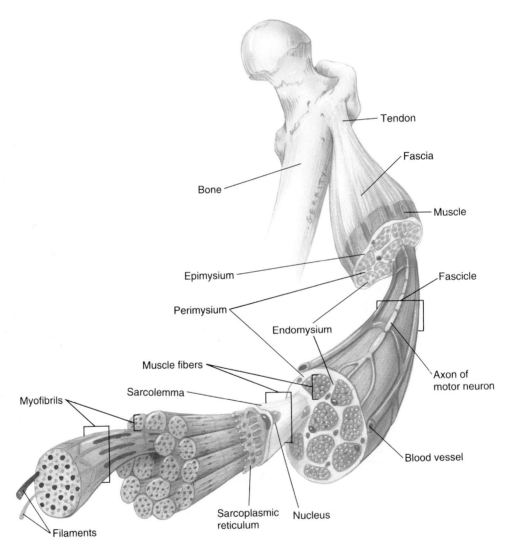

Tendon

Fascia

Bone

Muscle

Epimysium

Fascicle

Perimysium

Endomysium

Muscle fibers

Axon of
motor neuron

Sarcolemma

Myofibrils

Blood vessel

Filaments

Sarcoplasmic
reticulum

Nucleus

Skeletal Muscle Structure
Figure 8.1

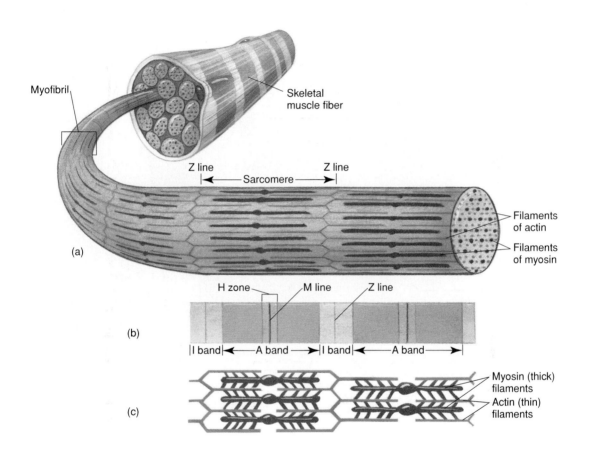

Myofibril

Skeletal muscle fiber

Z line   Z line
|←——— Sarcomere ———→|

Filaments of actin

Filaments of myosin

(a)

H zone   M line   Z line

(b)

|I band|←—— A band ——→|I band|←—— A band ——→|

Myosin (thick) filaments

Actin (thin) filaments

(c)

Skeletal Muscle Fiber I
Figure 8.2

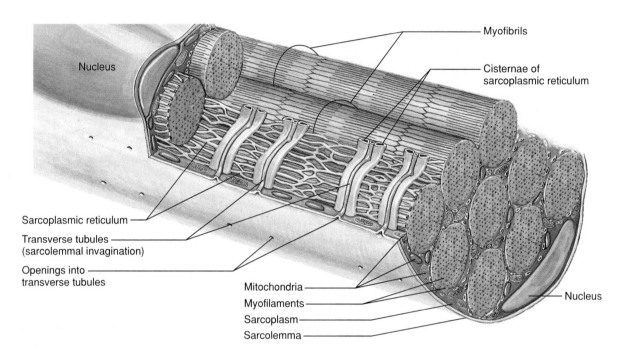

Nucleus

Myofibrils

Cisternae of
sarcoplasmic reticulum

Sarcoplasmic reticulum

Transverse tubules
(sarcolemmal invagination)

Openings into
transverse tubules

Mitochondria

Myofilaments

Sarcoplasm

Sarcolemma

Nucleus

Skeletal Muscle Fiber II
Figure 8.4

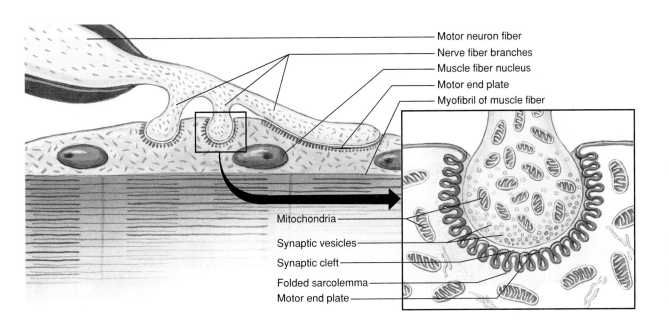

Motor neuron fiber
Nerve fiber branches
Muscle fiber nucleus
Motor end plate
Myofibril of muscle fiber

Mitochondria
Synaptic vesicles
Synaptic cleft
Folded sarcolemma
Motor end plate

Neuromuscular Junction
Figure 8.5

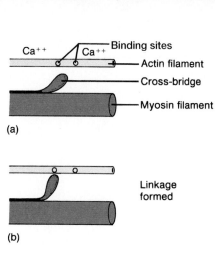

Ca^{++} Ca^{++} — Binding sites
— Actin filament
— Cross-bridge
— Myosin filament

(a)

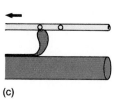

Linkage formed

(b)

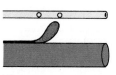

Cross-bridge pulling actin filament

(c)

Linkage broken

(d)

New linkage formed

(e)

Sliding Filament Theory
Figure 8.7

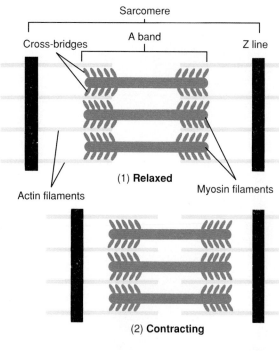

Sarcomere

Cross-bridges   A band   Z line

(1) **Relaxed**

Actin filaments   Myosin filaments

(2) **Contracting**

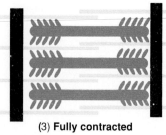

(3) **Fully contracted**

Contraction of a Sarcomere
Figure 8.8

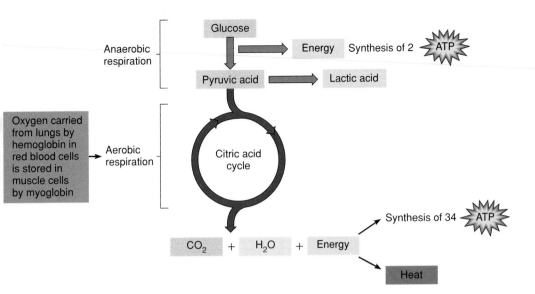

Muscle Metabolism
Figure 8.10

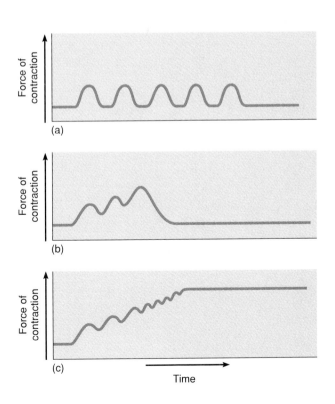

Muscle Fiber Contraction
Figure 8.12

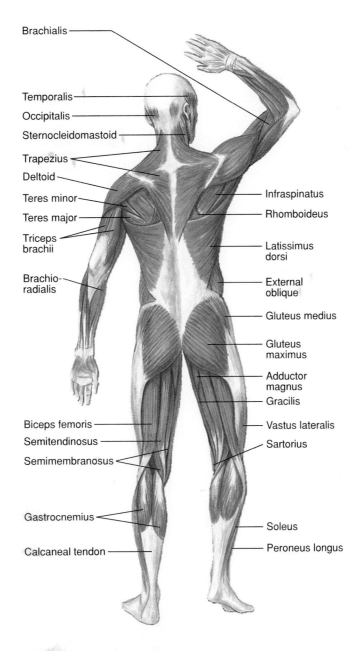

Brachialis

Temporalis

Occipitalis

Sternocleidomastoid

Trapezius

Deltoid

Teres minor

Teres major

Triceps
brachii

Brachio-
radialis

Infraspinatus

Rhomboideus

Latissimus
dorsi

External
oblique

Gluteus medius

Gluteus
maximus

Adductor
magnus

Gracilis

Biceps femoris

Semitendinosus

Semimembranosus

Vastus lateralis

Sartorius

Gastrocnemius

Calcaneal tendon

Soleus

Peroneus longus

Skeletal Muscles Posterior
Figure 8.15

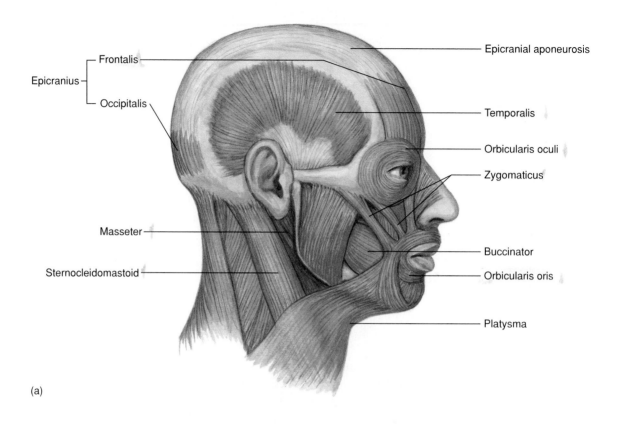

(a)

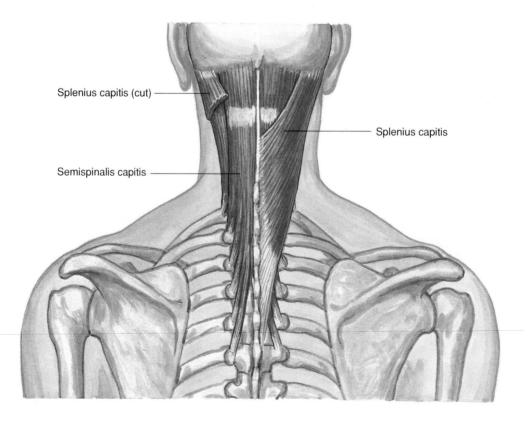

(b)

Muscles of Expression, Mastication and Head Movements
Figure 8.16a,b

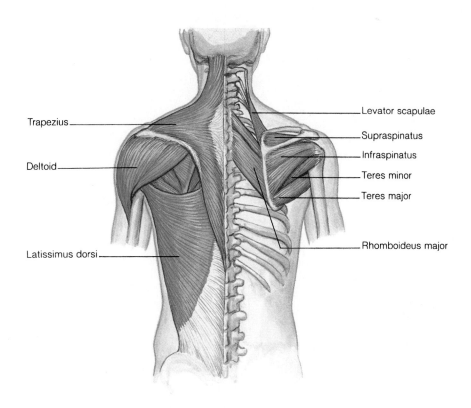

Trapezius

Deltoid

Latissimus dorsi

Levator scapulae

Supraspinatus

Infraspinatus

Teres minor

Teres major

Rhomboideus major

Muscles of the Posterior Shoulder
Figure 8.17

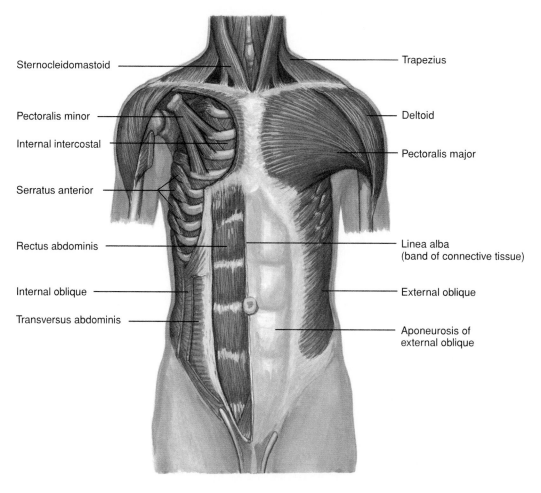

Sternocleidomastoid

Pectoralis minor

Internal intercostal

Serratus anterior

Rectus abdominis

Internal oblique

Transversus abdominis

Trapezius

Deltoid

Pectoralis major

Linea alba
(band of connective tissue)

External oblique

Aponeurosis of
external oblique

Muscles of the Anterior Chest and Abdominal Wall
Figure 8.18

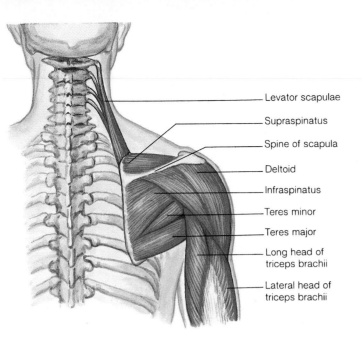

Levator scapulae

Supraspinatus

Spine of scapula

Deltoid

Infraspinatus

Teres minor

Teres major

Long head of
triceps brachii

Lateral head of
triceps brachii

Muscles of the Scapula and Arm
Figure 8.19

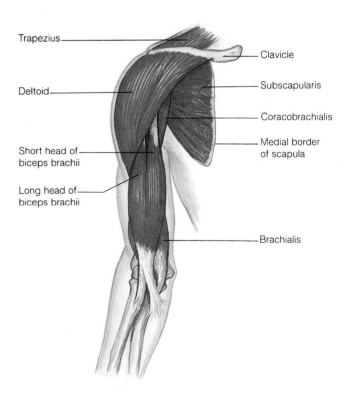

Trapezius

Clavicle

Deltoid

Subscapularis

Coracobrachialis

Medial border
of scapula

Short head of
biceps brachii

Long head of
biceps brachii

Brachialis

Muscles of the Anterior Shoulder and Arm
Figure 8.20

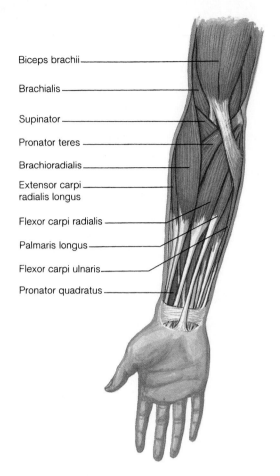

Biceps brachii

Brachialis

Supinator

Pronator teres

Brachioradialis

Extensor carpi radialis longus

Flexor carpi radialis

Palmaris longus

Flexor carpi ulnaris

Pronator quadratus

Muscles of the Anterior Forearm
Figure 8.21

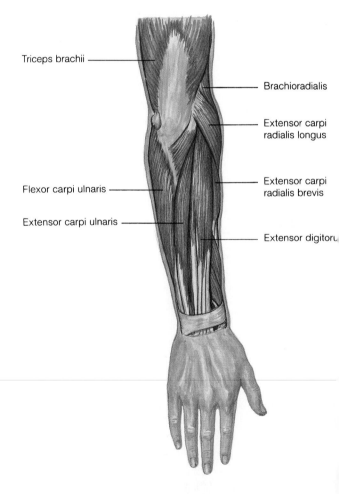

Triceps brachii

Brachioradialis

Extensor carpi radialis longus

Extensor carpi radialis brevis

Flexor carpi ulnaris

Extensor carpi ulnaris

Extensor digitoru

Muscles of the Posterior Forearm
Figure 8.22

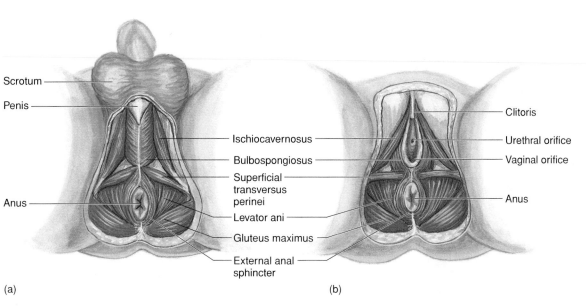

Scrotum

Penis

Anus

Ischiocavernosus

Bulbospongiosus

Superficial
transversus
perinei

Levator ani

Gluteus maximus

External anal
sphincter

Clitoris

Urethral orifice

Vaginal orifice

Anus

(a)

(b)

Muscles of the Pelvic Outlet
Figure 8.23

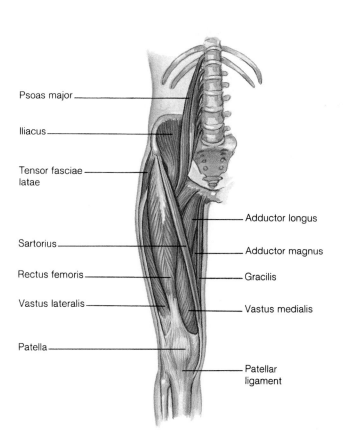

Psoas major

Iliacus

Tensor fasciae
latae

Sartorius

Rectus femoris

Vastus lateralis

Patella

Adductor longus

Adductor magnus

Gracilis

Vastus medialis

Patellar
ligament

Muscles of the Anterior Thigh
Figure 8.24

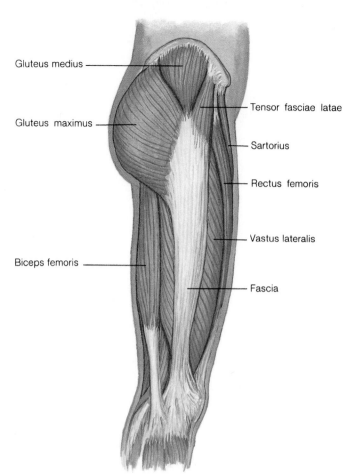

Gluteus medius

Gluteus maximus

Biceps femoris

Tensor fasciae latae

Sartorius

Rectus femoris

Vastus lateralis

Fascia

Muscles of the Lateral Thigh
Figure 8.25

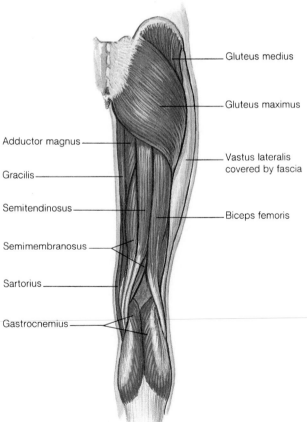

Gluteus medius

Gluteus maximus

Vastus lateralis covered by fascia

Biceps femoris

Adductor magnus

Gracilis

Semitendinosus

Semimembranosus

Sartorius

Gastrocnemius

Muscles of the Posterior Thigh
Figure 8.26

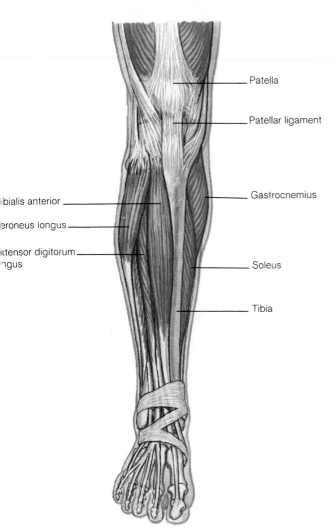

Patella

Patellar ligament

Gastrocnemius

ibialis anterior

eroneus longus

xtensor digitorum
ngus

Soleus

Tibia

Muscles of the Anterior Leg
Figure 8.27

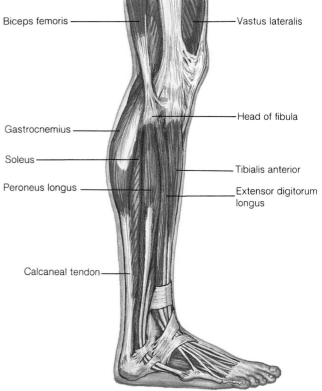

Biceps femoris

Vastus lateralis

Head of fibula

Gastrocnemius

Soleus

Tibialis anterior

Peroneus longus

Extensor digitorum
longus

Calcaneal tendon

Muscles of the Lateral Leg
Figure 8.28

71

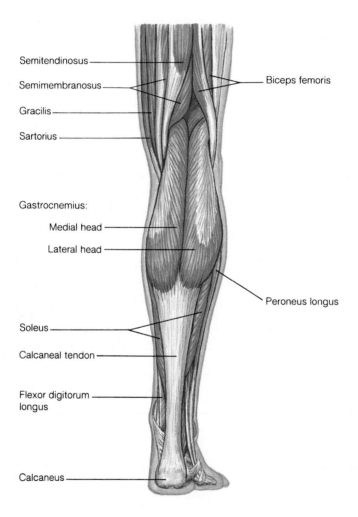

Semitendinosus

Semimembranosus

Gracilis

Sartorius

Biceps femoris

Gastrocnemius:

Medial head

Lateral head

Peroneus longus

Soleus

Calcaneal tendon

Flexor digitorum longus

Calcaneus

Muscles of the Posterior Leg
Figure 8.29

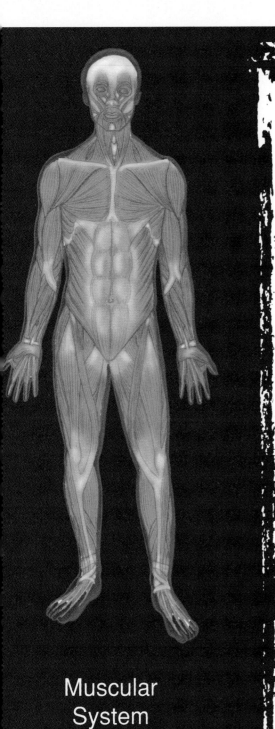

Muscular System

Muscles provide the force for moving body parts.

Integumentary System

The skin increases heat loss during skeletal muscle activity. Sensory receptors function in the reflex control of skeletal muscles.

Lymphatic System

Muscle action pumps lymph through lymphatic vessels.

Skeletal System

Bones provide attachments that allow skeletal muscles to cause movement.

Digestive System

Skeletal muscles are important in swallowing. The digestive system absorbs needed nutrients.

Nervous System

Neurons control muscle contractions.

Respiratory System

Breathing depends on skeletal muscles. The lungs provide oxygen for body cells and eliminate carbon dioxide.

Endocrine System

Hormones help increase blood flow to exercising skeletal muscles.

Urinary System

Skeletal muscles help control urine elimination.

Cardiovascular System

Blood flow delivers oxygen and nutrients and removes wastes.

Reproductive System

Skeletal muscles are important in sexual activity.

Muscular System
ORGANIZATION Chapter 8

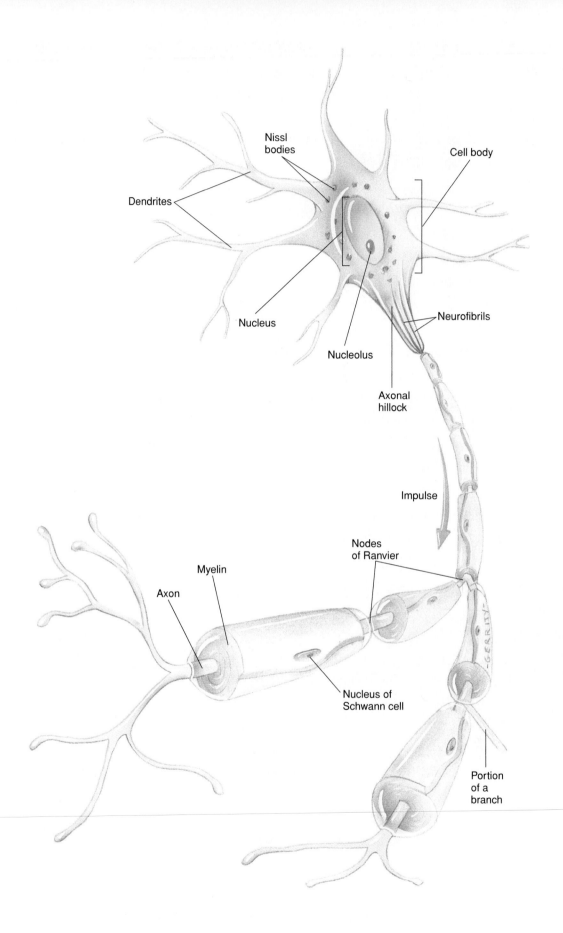

Nissl
bodies

Cell body

Dendrites

Nucleus

Nucleolus

Neurofibrils

Axonal
hillock

Impulse

Nodes
of Ranvier

Myelin

Axon

Nucleus of
Schwann cell

Portion
of a
branch

A Common Neuron
Figure 9.2

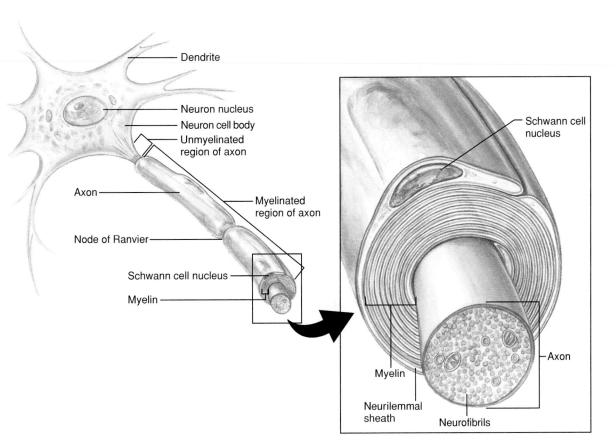

Schwann Cell
Figure 9.3

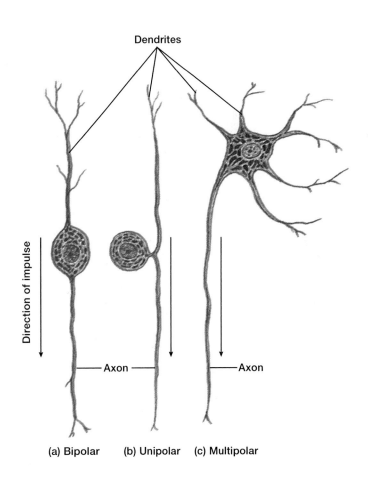

(a) Bipolar   (b) Unipolar   (c) Multipolar

Types of Neurons
Figure 9.4

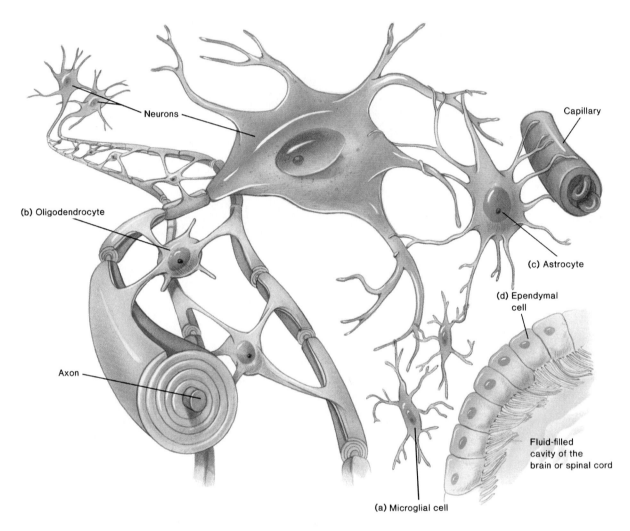

Neurons

Capillary

(b) Oligodendrocyte

(c) Astrocyte

(d) Ependymal
cell

Axon

Fluid-filled
cavity of the
brain or spinal cord

(a) Microglial cell

Neuroglial Cells
Figure 9.5

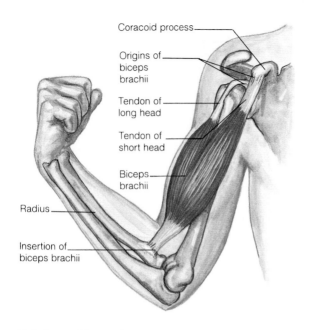

Origin and Insertion
Figure 8.13

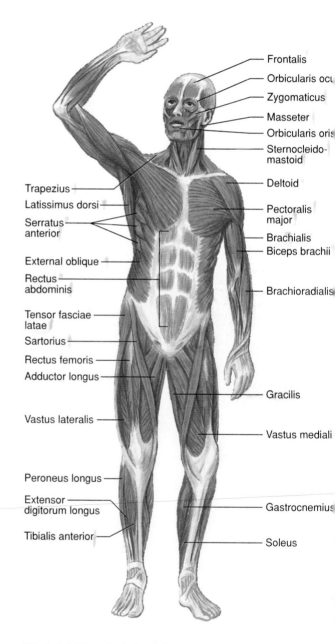

Coracoid process

Origins of biceps brachii

Tendon of long head

Tendon of short head

Biceps brachii

Radius

Insertion of biceps brachii

Trapezius

Latissimus dorsi

Serratus anterior

External oblique

Rectus abdominis

Tensor fasciae latae

Sartorius

Rectus femoris

Adductor longus

Vastus lateralis

Peroneus longus

Extensor digitorum longus

Tibialis anterior

Frontalis

Orbicularis ocu

Zygomaticus

Masseter

Orbicularis oris

Sternocleido-mastoid

Deltoid

Pectoralis major

Brachialis

Biceps brachii

Brachioradialis

Gracilis

Vastus mediali

Gastrocnemius

Soleus

Skeletal Muscles Anterior
Figure 8.14

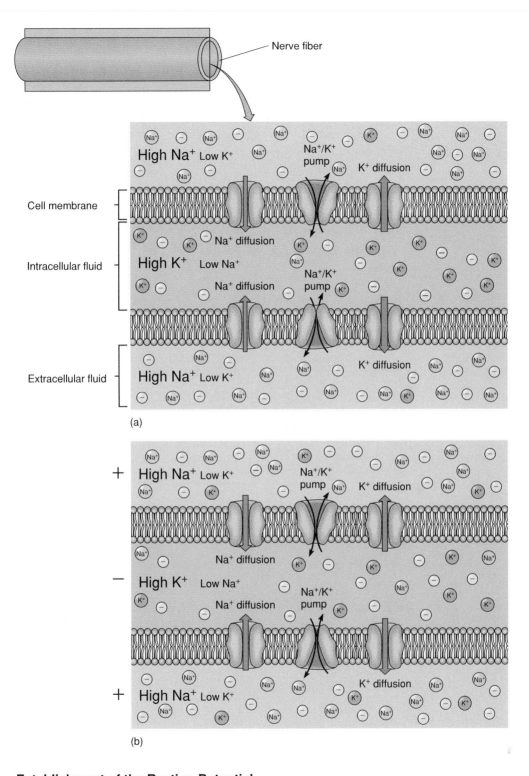

Establishment of the Resting Potential
Figure 9.7

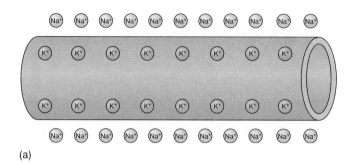

(a)

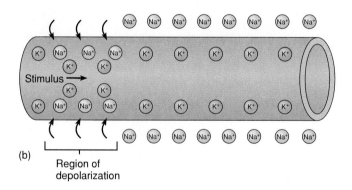

(b)

Region of
depolarization

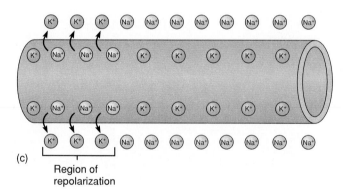

(c)

Region of
repolarization

Action Potential
Figure 9.8

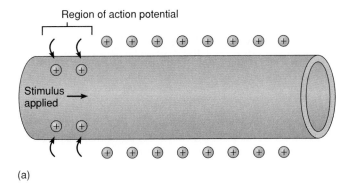

(a)

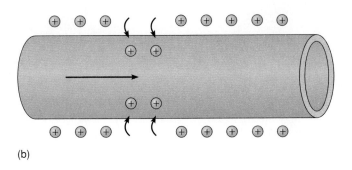

(b)

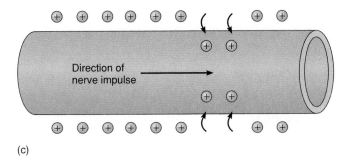

(c)

Impulse Conduction
Figure 9.9

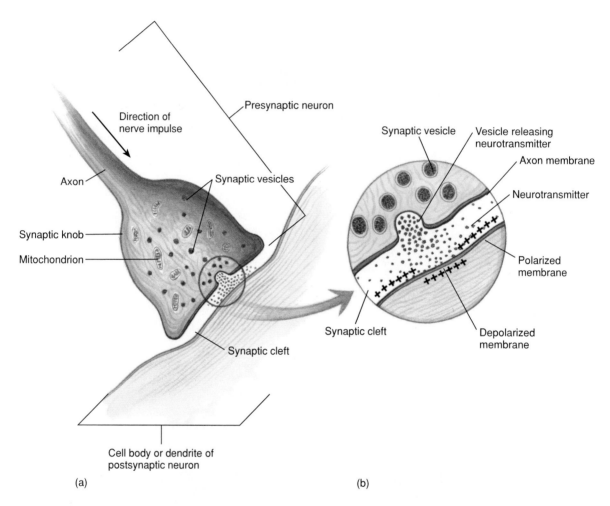

Direction of
nerve impulse

Presynaptic neuron

Axon

Synaptic vesicles

Synaptic knob

Mitochondrion

Synaptic cleft

Cell body or dendrite of
postsynaptic neuron

(a)

Synaptic vesicle

Vesicle releasing
neurotransmitter

Axon membrane

Neurotransmitter

Polarized
membrane

Synaptic cleft

Depolarized
membrane

(b)

Synaptic Knob and Synaptic Cleft
Figure 9.11

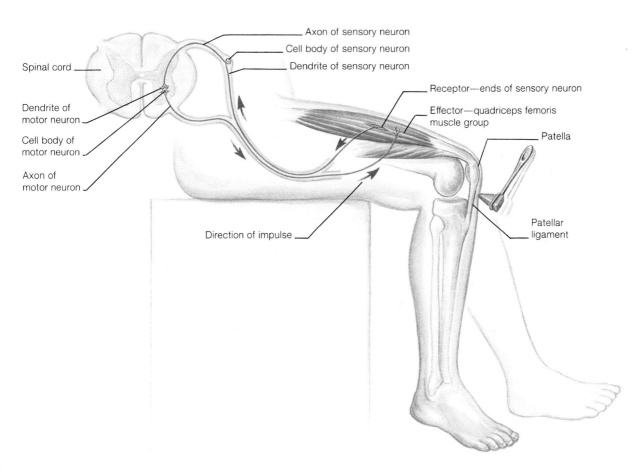

Spinal cord

Axon of sensory neuron

Cell body of sensory neuron

Dendrite of sensory neuron

Receptor—ends of sensory neuron

Dendrite of
motor neuron

Effector—quadriceps femoris
muscle group

Cell body of
motor neuron

Patella

Axon of
motor neuron

Direction of impulse

Patellar
ligament

Knee-Jerk Reflex
Figure 9.15

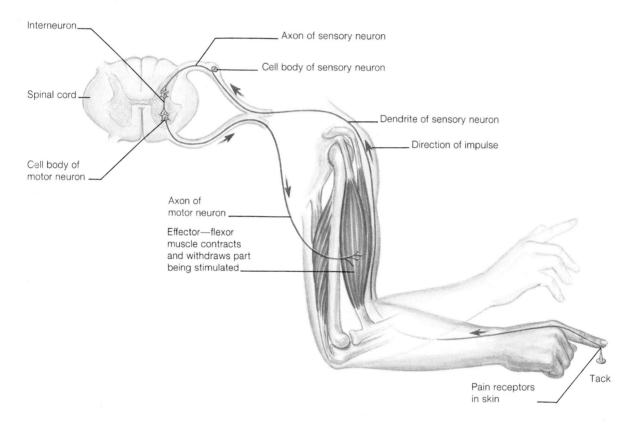

Interneuron

Axon of sensory neuron

Cell body of sensory neuron

Spinal cord

Dendrite of sensory neuron

Direction of impulse

Cell body of
motor neuron

Axon of
motor neuron

Effector—flexor
muscle contracts
and withdraws part
being stimulated

Tack

Pain receptors
in skin

Withdrawal Reflex
Figure 9.16

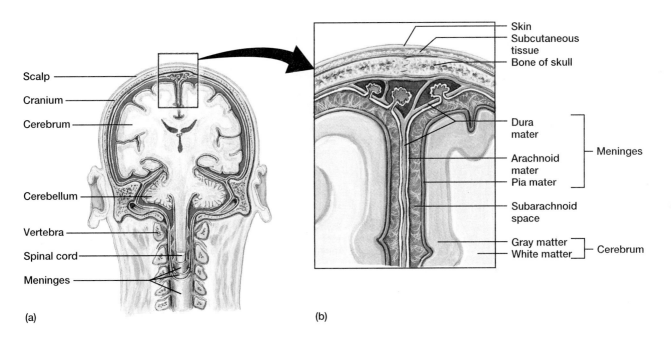

Scalp

Cranium

Cerebrum

Cerebellum

Vertebra

Spinal cord

Meninges

(a)

Skin
Subcutaneous
tissue
Bone of skull

Dura
mater

Arachnoid
mater
Pia mater

Meninges

Subarachnoid
space

Gray matter
White matter

Cerebrum

(b)

Meninges I
Figure 9.17

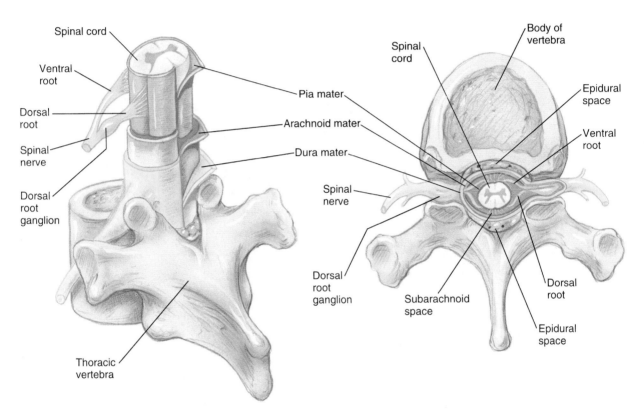

Spinal cord

Ventral root

Dorsal root

Spinal nerve

Dorsal root ganglion

Pia mater

Arachnoid mater

Dura mater

Thoracic vertebra

Body of vertebra

Spinal cord

Epidural space

Ventral root

Spinal nerve

Dorsal root ganglion

Subarachnoid space

Dorsal root

Epidural space

Meninges II
Figure 9.18

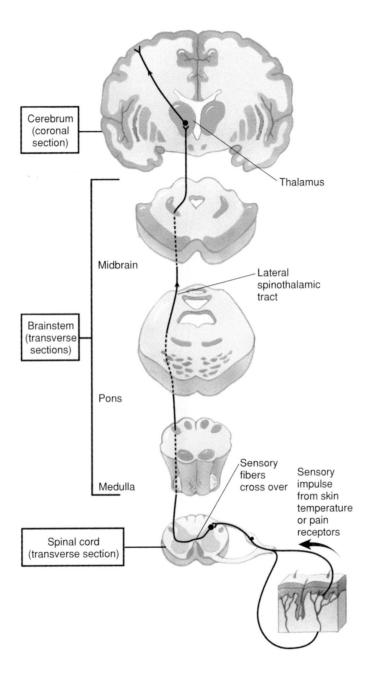

Cerebrum (coronal section)

Thalamus

Midbrain

Lateral spinothalamic tract

Brainstem (transverse sections)

Pons

Medulla

Sensory fibers cross over

Sensory impulse from skin temperature or pain receptors

Spinal cord (transverse section)

Ascending Tracts
Figure 9.21

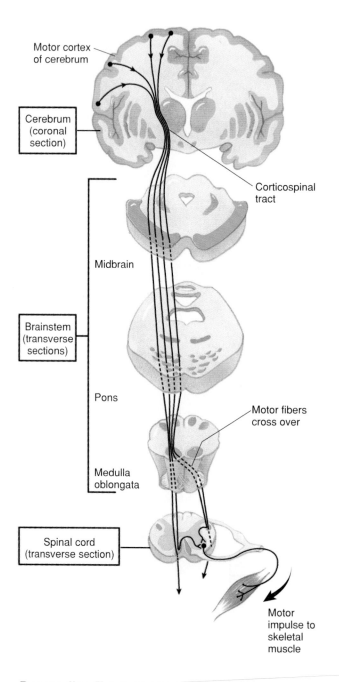

Motor cortex
of cerebrum

Cerebrum
(coronal
section)

Corticospinal
tract

Midbrain

Brainstem
(transverse
sections)

Pons

Motor fibers
cross over

Medulla
oblongata

Spinal cord
(transverse section)

Motor
impulse to
skeletal
muscle

Descending Tracts
Figure 9.22

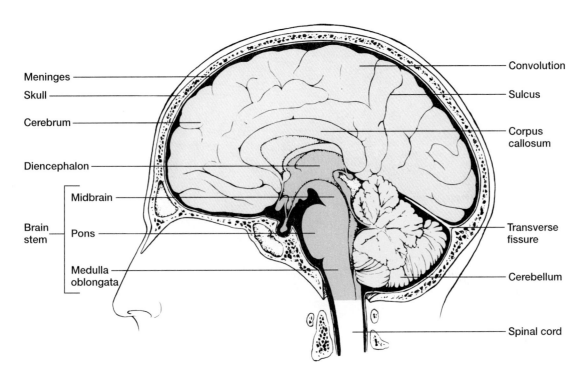

Meninges

Skull

Cerebrum

Diencephalon

Midbrain

Brain
stem

Pons

Medulla
oblongata

Convolution

Sulcus

Corpus
callosum

Transverse
fissure

Cerebellum

Spinal cord

The Brain
Figure 9.23

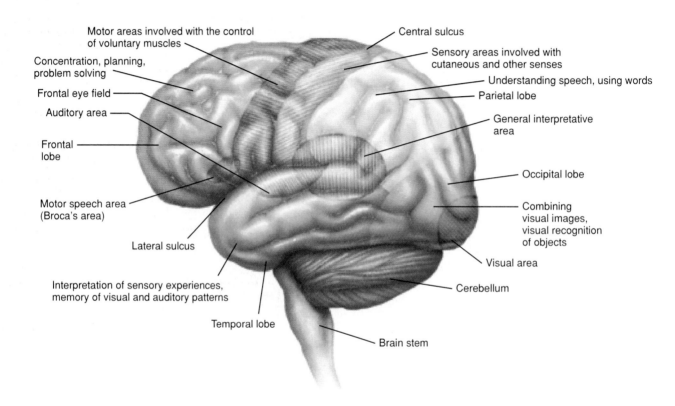

Motor areas involved with the control of voluntary muscles

Concentration, planning, problem solving

Frontal eye field

Auditory area

Frontal lobe

Motor speech area (Broca's area)

Lateral sulcus

Interpretation of sensory experiences, memory of visual and auditory patterns

Temporal lobe

Central sulcus

Sensory areas involved with cutaneous and other senses

Understanding speech, using words

Parietal lobe

General interpretative area

Occipital lobe

Combining visual images, visual recognition of objects

Visual area

Cerebellum

Brain stem

Sensory, Motor and Association Areas
Figure 9.24

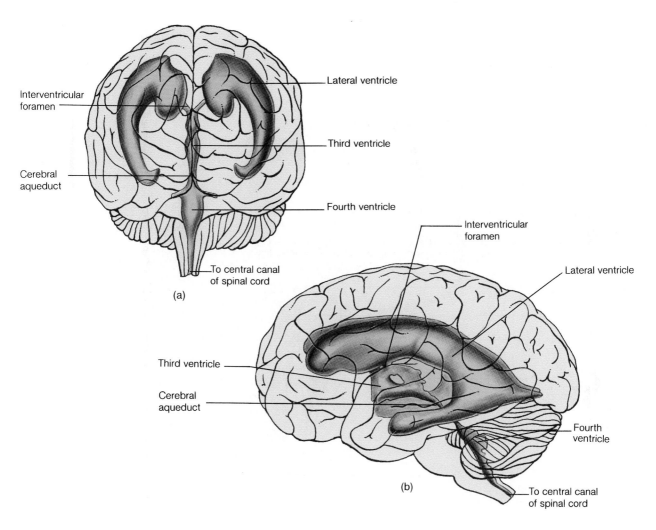

Interventricular foramen

Lateral ventricle

Cerebral aqueduct

Third ventricle

Fourth ventricle

To central canal of spinal cord

(a)

Interventricular foramen

Lateral ventricle

Third ventricle

Cerebral aqueduct

Fourth ventricle

To central canal of spinal cord

(b)

Ventricles of the Brain
Figure 9.26

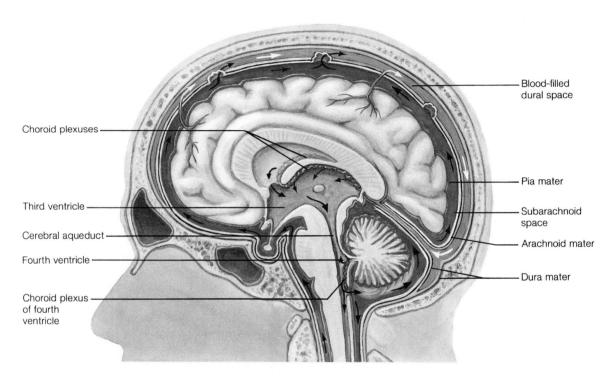

Choroid plexuses

Third ventricle

Cerebral aqueduct

Fourth ventricle

Choroid plexus
of fourth
ventricle

Blood-filled
dural space

Pia mater

Subarachnoid
space

Arachnoid mater

Dura mater

Cerebrospinal Fluid Circulation
Figure 9.27

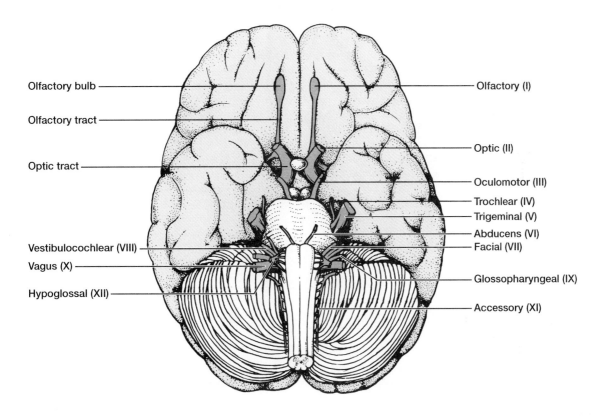

Olfactory bulb

Olfactory tract

Optic tract

Vestibulocochlear (VIII)

Vagus (X)

Hypoglossal (XII)

Olfactory (I)

Optic (II)

Oculomotor (III)

Trochlear (IV)

Trigeminal (V)

Abducens (VI)

Facial (VII)

Glossopharyngeal (IX)

Accessory (XI)

Cranial Nerves
Figure 9.30

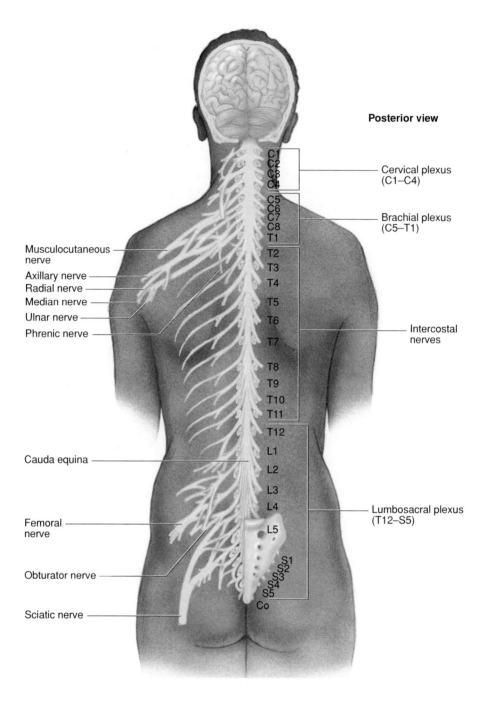

Posterior view

C1
C2
C3
C4 — Cervical plexus (C1–C4)

C5
C6
C7
C8
T1 — Brachial plexus (C5–T1)

Musculocutaneous nerve
Axillary nerve
Radial nerve
Median nerve
Ulnar nerve
Phrenic nerve

T2
T3
T4
T5
T6
T7 — Intercostal nerves

T8
T9
T10
T11

T12
L1
L2
L3
L4
L5 — Lumbosacral plexus (T12–S5)

Cauda equina

Femoral nerve

Obturator nerve

Sciatic nerve

S1
S2
S3
S4
S5
Co

Spinal Nerves
Figure 9.31

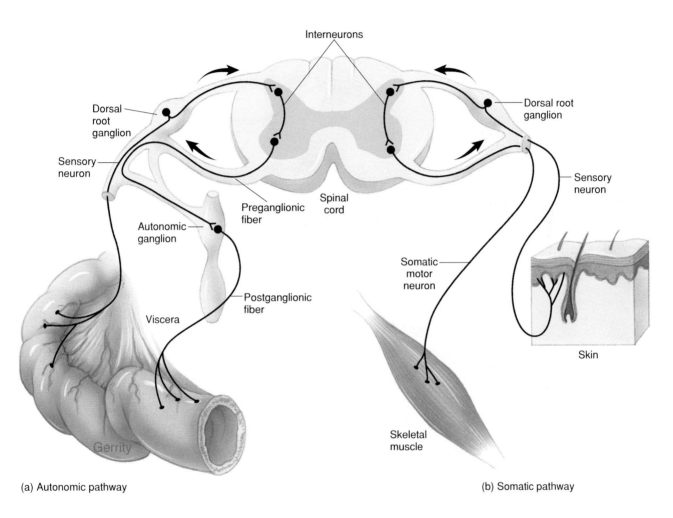

Interneurons

Dorsal
root
ganglion

Dorsal root
ganglion

Sensory
neuron

Preganglionic
fiber

Spinal
cord

Sensory
neuron

Autonomic
ganglion

Somatic
motor
neuron

Postganglionic
fiber

Viscera

Skin

Gerrity

Skeletal
muscle

(a) Autonomic pathway

(b) Somatic pathway

Somatic and Autonomic Nerve Pathways
Figure 9.32

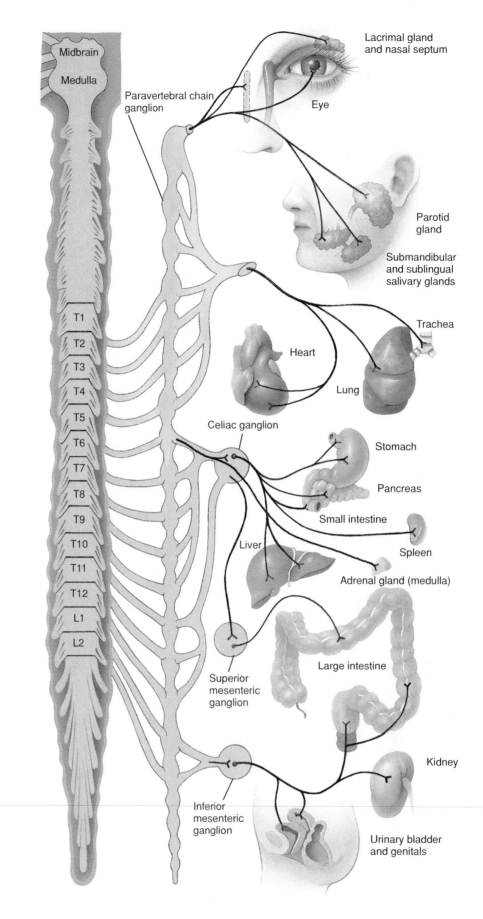

Sympathetic Nervous System
Figure 9.33

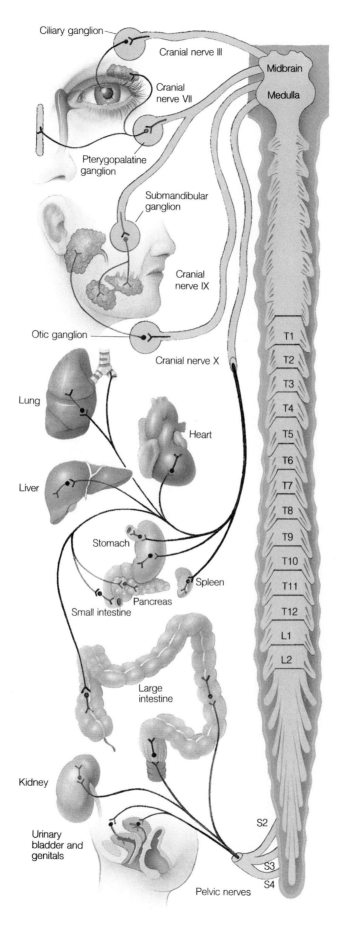

Parasympathetic Nervous System
Figure 9.34

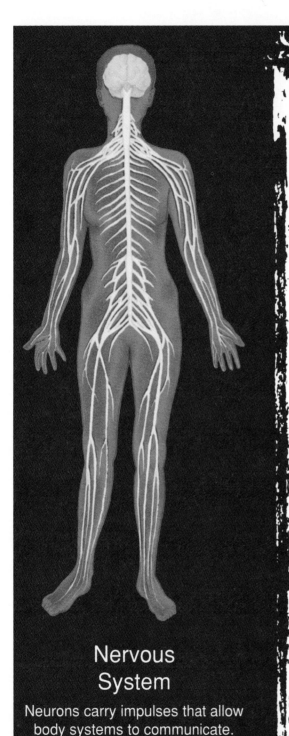

Nervous System

Neurons carry impulses that allow body systems to communicate.

Integumentary System

Sensory receptors provide the nervous system with information about the outside world.

Skeletal System

Bones protect the brain and spinal cord and help maintain plasma calcium, which is important to neuron function.

Muscular System

Nerve impulses control movement and carry information about the position of body parts.

Endocrine System

The hypothalamus controls secretion of many hormones.

Cardiovascular System

Nerve impulses help control blood flow and blood pressure.

Lymphatic System

Stress may impair the immune response.

Digestive System

The nervous system can influence digestive function.

Respiratory System

The nervous system alters respiratory activity to control oxygen levels and blood pH.

Urinary System

Nerve impulses affect urine production and elimination.

Reproductive System

The nervous system plays a role in egg and sperm formation, sexual pleasure, childbirth, and nursing.

Nervous System
ORGANIZATION Chapter 9

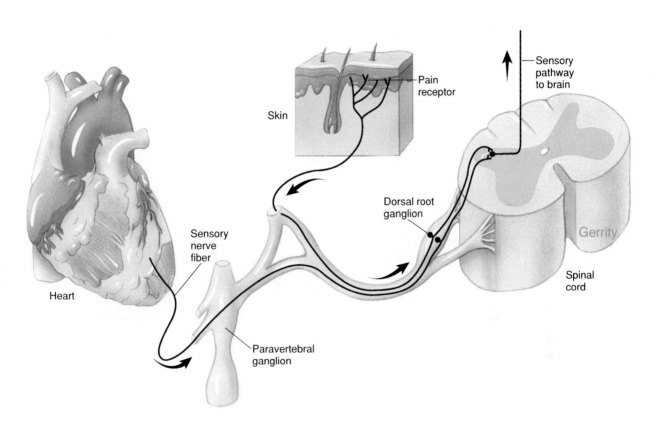

Referred Pain Pathway
Figure 10.3

Labels in figure: Skin, Pain receptor, Sensory pathway to brain, Dorsal root ganglion, Spinal cord, Sensory nerve fiber, Heart, Paravertebral ganglion, Gerrity

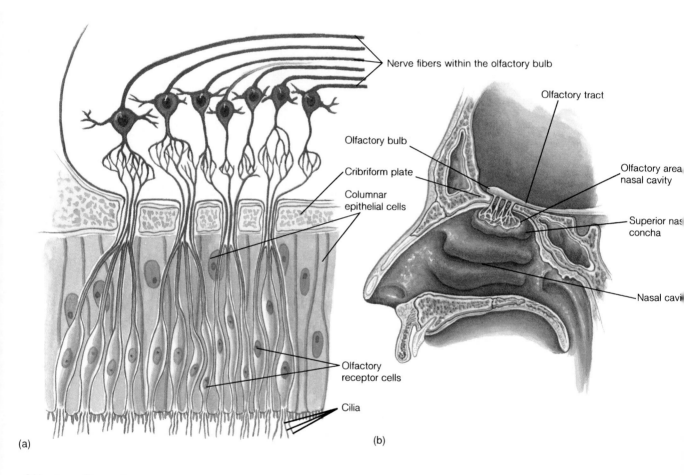

Nerve fibers within the olfactory bulb

Olfactory tract

Olfactory bulb

Olfactory area
nasal cavity

Cribriform plate

Columnar
epithelial cells

Superior nas
concha

Olfactory
receptor cells

Nasal cavi

Cilia

(a)

(b)

Olfactory Receptors
Figure 10.4

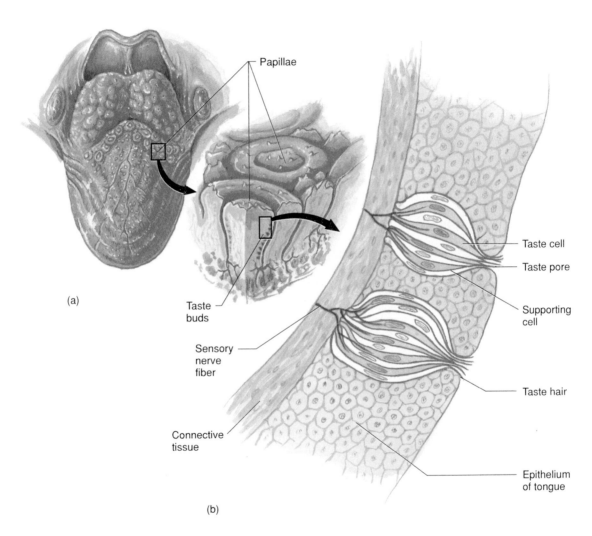

Papillae

Taste cell

Taste pore

Supporting cell

Taste hair

Epithelium of tongue

(a)

Taste buds

Sensory nerve fiber

Connective tissue

(b)

Taste Receptors
Figure 10.5

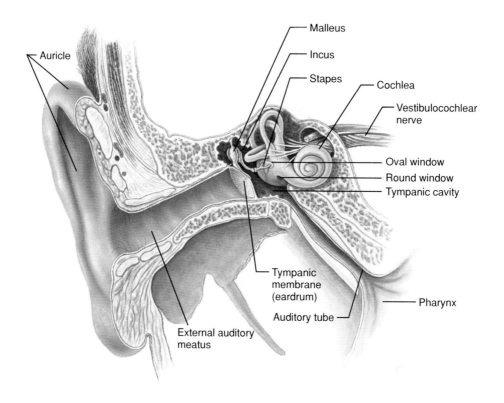

Major Parts of the Ear
Figure 10.7

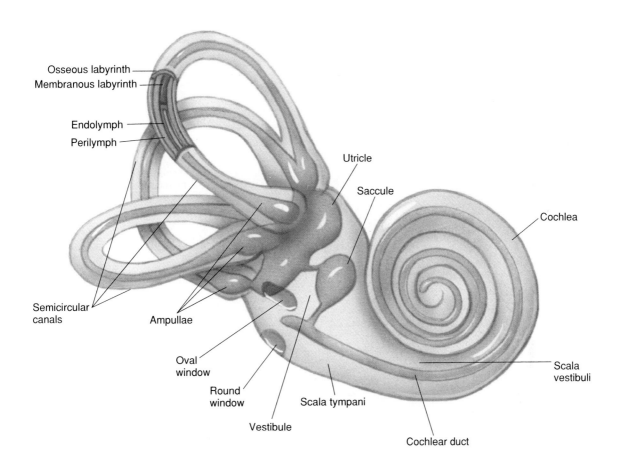

Osseous labyrinth

Membranous labyrinth

Endolymph

Perilymph

Utricle

Saccule

Cochlea

Semicircular
canals

Ampullae

Oval
window

Round
window

Vestibule

Scala tympani

Cochlear duct

Scala
vestibuli

Inner Ear Structure
Figure 10.9

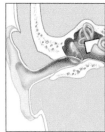

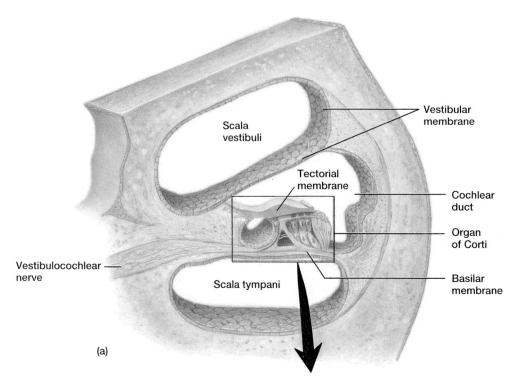

Scala
vestibuli

Vestibular
membrane

Tectorial
membrane

Cochlear
duct

Organ
of Corti

Vestibulocochlear
nerve

Scala tympani

Basilar
membrane

(a)

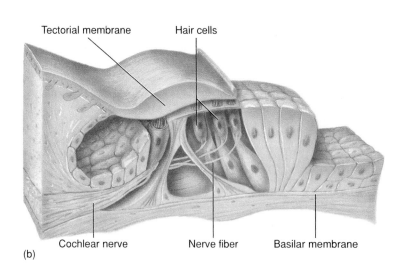

Tectorial membrane

Hair cells

Cochlear nerve

Nerve fiber

Basilar membrane

(b)

Organ of Corti
Figure 10.10

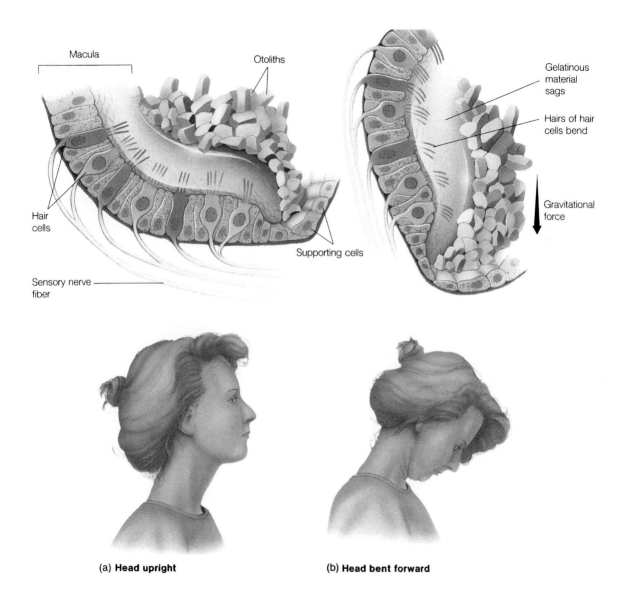

Macula

Otoliths

Gelatinous material sags

Hairs of hair cells bend

Gravitational force

Hair cells

Supporting cells

Sensory nerve fiber

(a) **Head upright**

(b) **Head bent forward**

Static Equilibrium
Figure 10.12

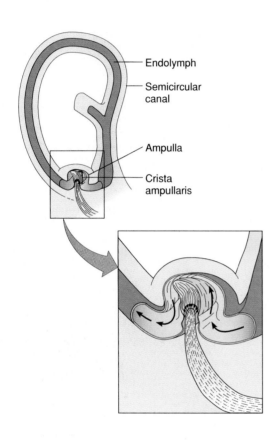

Endolymph

Semicircular canal

Ampulla

Crista ampullaris

(a) Head in still position

(b) Head rotating

Dynamic Equilibrium
Figure 10.14

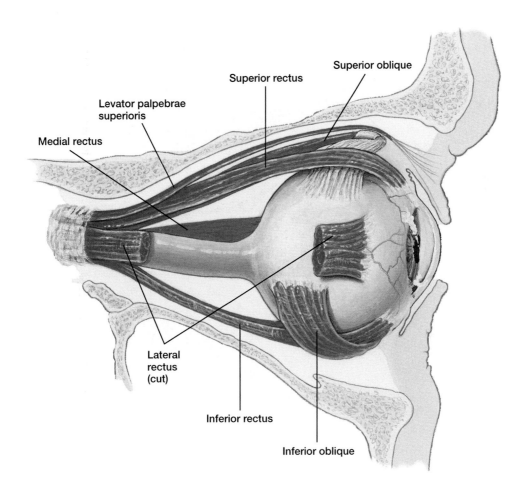

Extrinsic Muscles of the Eye
Figure 10.17

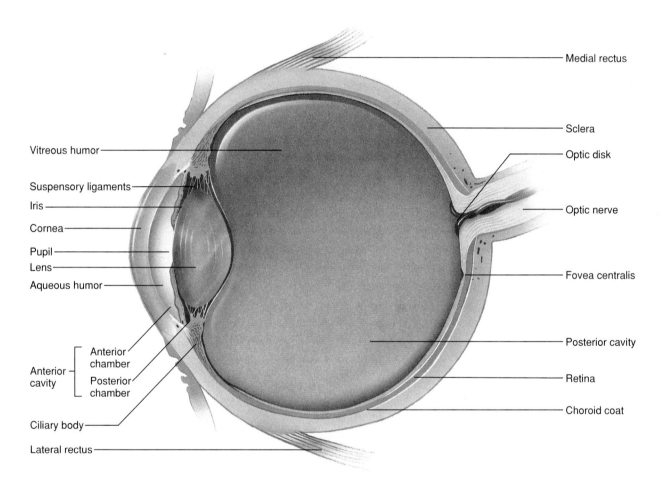

Vitreous humor

Suspensory ligaments

Iris

Cornea

Pupil

Lens

Aqueous humor

Anterior chamber

Anterior cavity

Posterior chamber

Ciliary body

Lateral rectus

Medial rectus

Sclera

Optic disk

Optic nerve

Fovea centralis

Posterior cavity

Retina

Choroid coat

Eye, Transverse Section
Figure 10.18

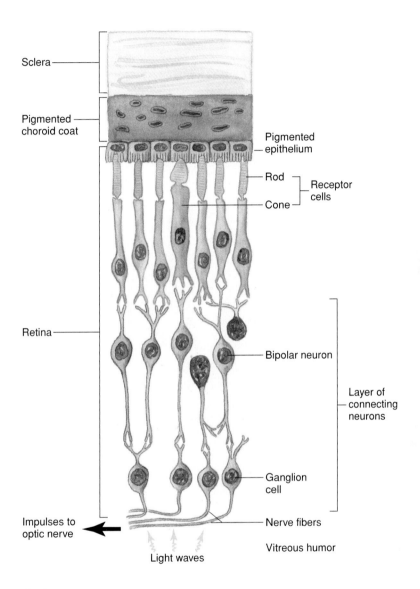

Sclera

Pigmented
choroid coat

Pigmented
epithelium

Rod

Cone

Receptor
cells

Retina

Bipolar neuron

Layer of
connecting
neurons

Ganglion
cell

Impulses to
optic nerve

Nerve fibers

Vitreous humor

Light waves

Retina
Figure 10.22

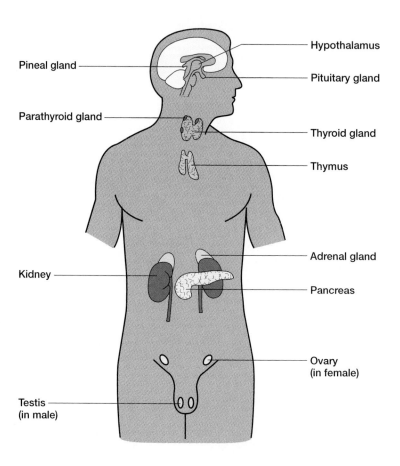

Pineal gland

Hypothalamus

Pituitary gland

Parathyroid gland

Thyroid gland

Thymus

Kidney

Adrenal gland

Pancreas

Ovary
(in female)

Testis
(in male)

Major Endocrine Glands
Figure 11.1

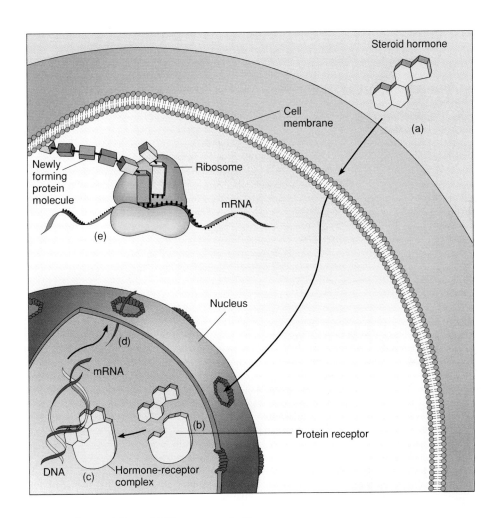

Mechanism of Steroid Hormone Action
Figure 11.2

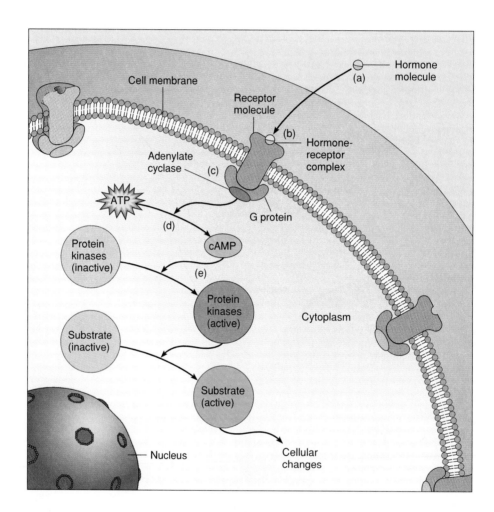

Mechanism of Non-Steroid Hormone Action
Figure 11.3

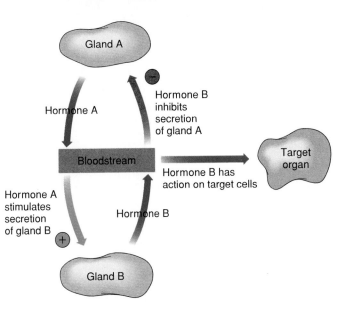

Anterior and Posterior Pituitary Glands
Figure 11.4

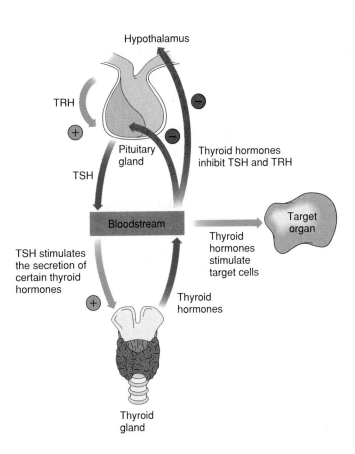

Control of Thyroid Hormone
Figure 11.8

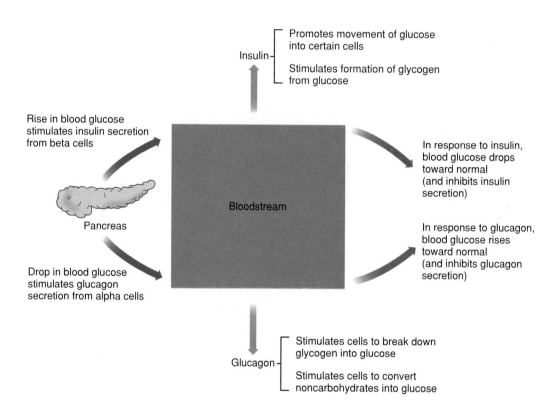

Insulin
- Promotes movement of glucose into certain cells
- Stimulates formation of glycogen from glucose

Rise in blood glucose stimulates insulin secretion from beta cells

In response to insulin, blood glucose drops toward normal (and inhibits insulin secretion)

Bloodstream

Pancreas

In response to glucagon, blood glucose rises toward normal (and inhibits glucagon secretion)

Drop in blood glucose stimulates glucagon secretion from alpha cells

Glucagon
- Stimulates cells to break down glycogen into glucose
- Stimulates cells to convert noncarbohydrates into glucose

Control of Insulin and Glucagon
Figure 11.16

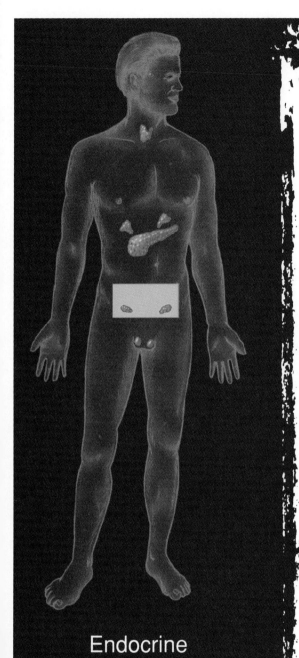

Endocrine System

Glands secrete hormones that have a variety of effects on cells, tissues, organs, and organ systems.

Integumentary System

Melanocytes produce skin pigment in response to hormonal stimulation.

Lymphatic System

Hormones stimulate lymphocyte production.

Skeletal System

Hormones act on bones to control calcium balance.

Digestive System

Hormones help control digestive system activity.

Muscular System

Hormones help increase blood flow to exercising muscles.

Respiratory System

Decreased oxygen causes hormonal stimulation of red blood cell production; red blood cells transport oxygen and carbon dioxide.

Nervous System

Neurons control the secretions of the anterior and posterior pituitary glands and the adrenal medulla.

Urinary System

Hormones act on the kidneys to help control water and electrolyte balance.

Cardiovascular System

Hormones are carried in the bloodstream; some have direct actions on the heart and blood vessels.

Reproductive System

Sex hormones play a major role in development of secondary sex characteristics, egg, and sperm.

Endocrine System
ORGANIZATION Chapter 11

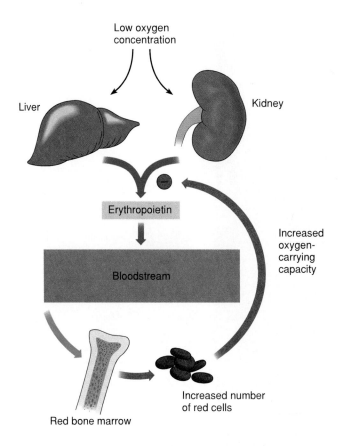

Stimulation of Red Blood Cell Production
Figure 12.3

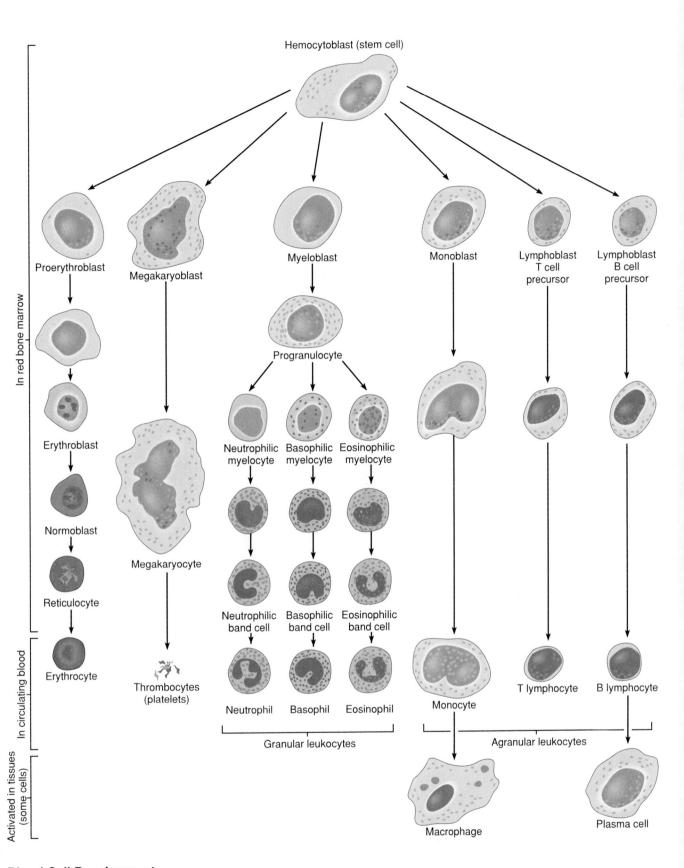

Hemocytoblast (stem cell)

In red bone marrow

Proerythroblast

Megakaryoblast

Myeloblast

Monoblast

Lymphoblast T cell precursor

Lymphoblast B cell precursor

Erythroblast

Progranulocyte

Normoblast

Neutrophilic myelocyte

Basophilic myelocyte

Eosinophilic myelocyte

Reticulocyte

Megakaryocyte

Neutrophilic band cell

Basophilic band cell

Eosinophilic band cell

In circulating blood

Erythrocyte

Thrombocytes (platelets)

Neutrophil

Basophil

Eosinophil

Granular leukocytes

Monocyte

T lymphocyte

B lymphocyte

Agranular leukocytes

Activated in tissues (some cells)

Macrophage

Plasma cell

Blood Cell Development
Figure 12.4

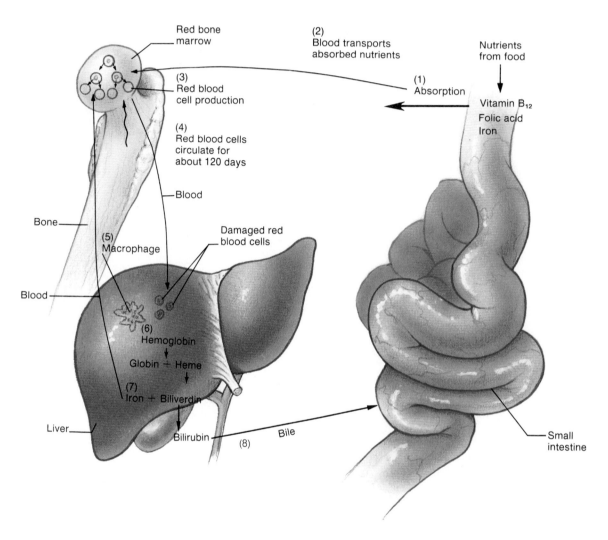

Red bone marrow

(2) Blood transports absorbed nutrients

Nutrients from food

(3) Red blood cell production

(1) Absorption

Vitamin B₁₂
Folic acid
Iron

(4) Red blood cells circulate for about 120 days

Blood

Bone

(5) Macrophage

Damaged red blood cells

Blood

(6) Hemoglobin

Globin + Heme

(7) Iron + Biliverdin

Liver

Bilirubin

(8) Bile

Small intestine

Life Cycle of Red Blood Cell
Figure 12.5

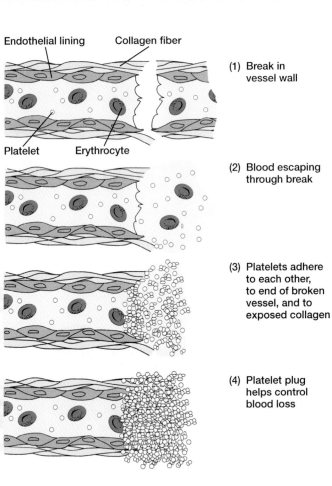

Endothelial lining   Collagen fiber

(1) Break in
    vessel wall

Platelet   Erythrocyte

(2) Blood escaping
    through break

(3) Platelets adhere
    to each other,
    to end of broken
    vessel, and to
    exposed collagen

(4) Platelet plug
    helps control
    blood loss

Platelet Plug Formation
Figure 12.12

TISSUE DAMAGE

Blood vessel
spasm

Platelet plug
formation

Clotting
mechanism

Prothrombin

Prothrombin
activator → Ca⁺⁺   Fibrinogen

Thrombin → Ca⁺⁺

Fibrin

Blood clot
formation

Hemostasis
Figure 12.15

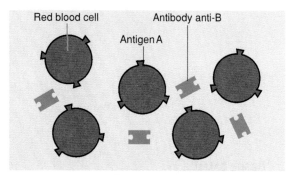

Type A blood

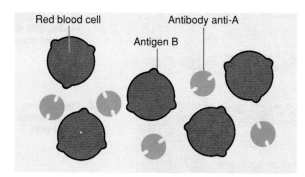

Type B blood

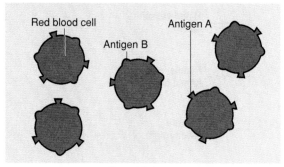

Type AB blood

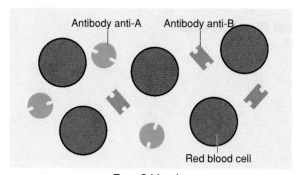

Type O blood

ABO Blood Types
Figure 12.16

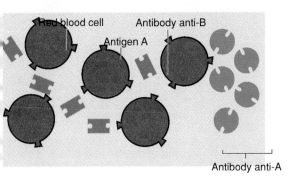

Red blood cell   Antibody anti-B
        Antigen A

Antibody anti-A

(a)

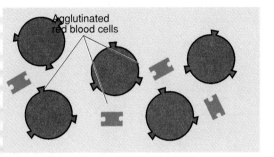

Agglutinated
red blood cells

(b)

ABO Blood Reactions
Figure 12.17

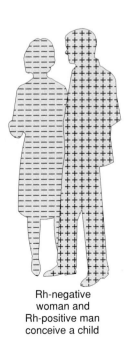

Rh-negative
woman and
Rh-positive man
conceive a child

Rh-negative
woman with
Rh-positive
fetus

Cells from
Rh-positive
fetus enter
mother's
bloodstream

Woman
becomes
sensitized—
antibodies (⊕)
form to fight
Rh-positive
blood cells

In the next
Rh-positive
pregnancy,
maternal
antibodies
attack fetal
blood cells

Rh Factor
Figure 12.18

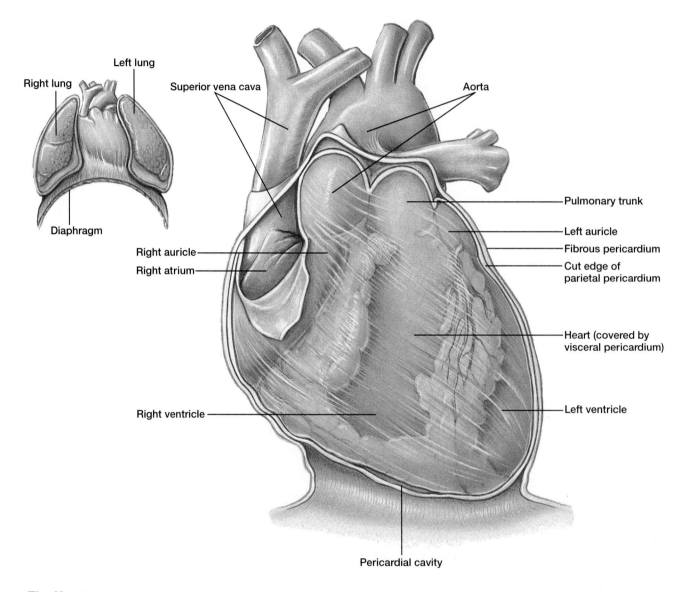

Right lung   Left lung

Superior vena cava

Aorta

Pulmonary trunk

Left auricle

Diaphragm

Fibrous pericardium

Right auricle

Cut edge of
parietal pericardium

Right atrium

Heart (covered by
visceral pericardium)

Right ventricle

Left ventricle

Pericardial cavity

The Heart
Figure 13.2

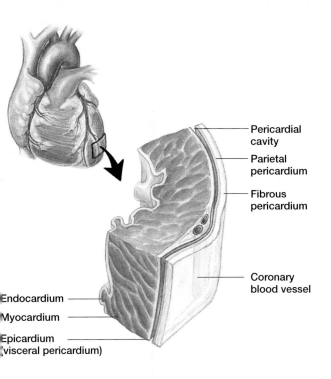

Pericardial cavity

Parietal pericardium

Fibrous pericardium

Coronary blood vessel

Endocardium

Myocardium

Epicardium (visceral pericardium)

Wall of the Heart
Figure 13.3

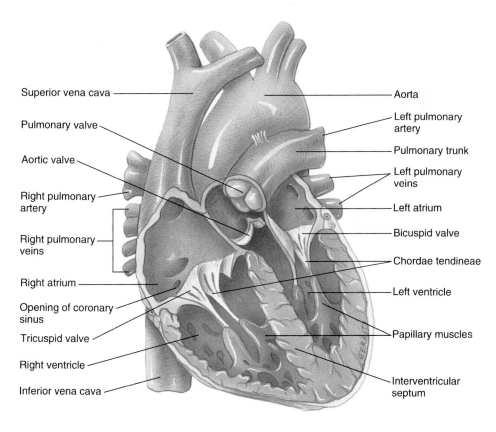

Superior vena cava

Pulmonary valve

Aortic valve

Right pulmonary artery

Right pulmonary veins

Right atrium

Opening of coronary sinus

Tricuspid valve

Right ventricle

Inferior vena cava

Aorta

Left pulmonary artery

Pulmonary trunk

Left pulmonary veins

Left atrium

Bicuspid valve

Chordae tendineae

Left ventricle

Papillary muscles

Interventricular septum

The Heart, Coronal Sections
Figure 13.4

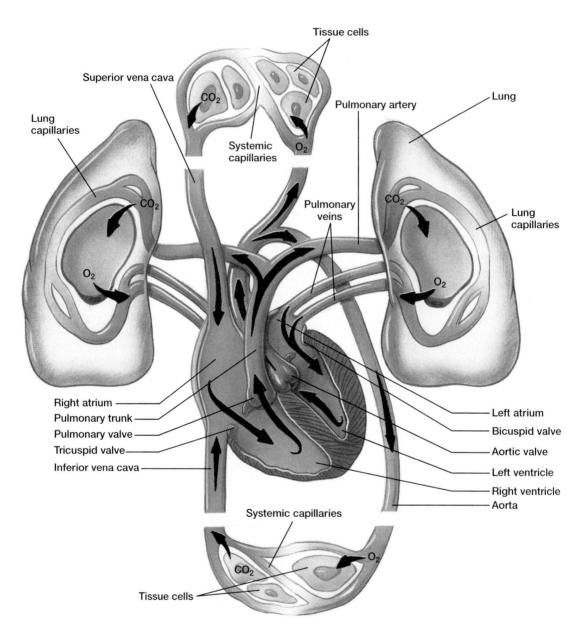

Blood Flow to and from the Heart
Figure 13.7

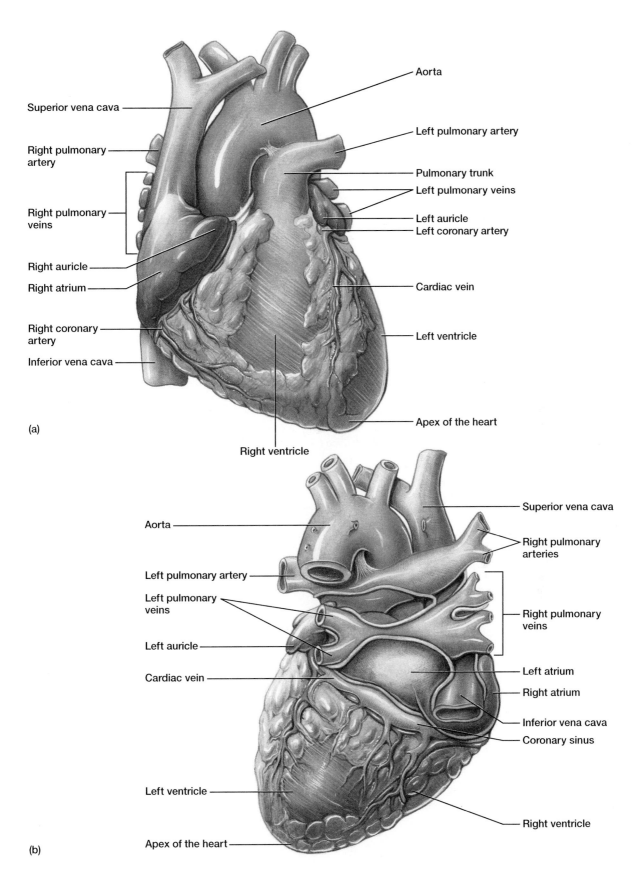

(a)

Superior vena cava

Right pulmonary artery

Right pulmonary veins

Right auricle

Right atrium

Right coronary artery

Inferior vena cava

Aorta

Left pulmonary artery

Pulmonary trunk

Left pulmonary veins

Left auricle

Left coronary artery

Cardiac vein

Left ventricle

Apex of the heart

Right ventricle

(b)

Aorta

Left pulmonary artery

Left pulmonary veins

Left auricle

Cardiac vein

Left ventricle

Apex of the heart

Superior vena cava

Right pulmonary arteries

Right pulmonary veins

Left atrium

Right atrium

Inferior vena cava

Coronary sinus

Right ventricle

The Heart and Coronary Vessels
Figure 13.9

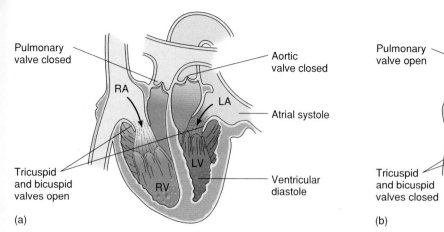

Pulmonary valve closed

Aortic valve closed

Atrial systole

RA

LA

Tricuspid and bicuspid valves open

LV

RV

Ventricular diastole

(a)

Pulmonary valve open

Aortic valve open

Atrial diast●

Tricuspid and bicuspid valves closed

Ventricula● systole

(b)

Atrial Systole and Diastole
Figure 13.10

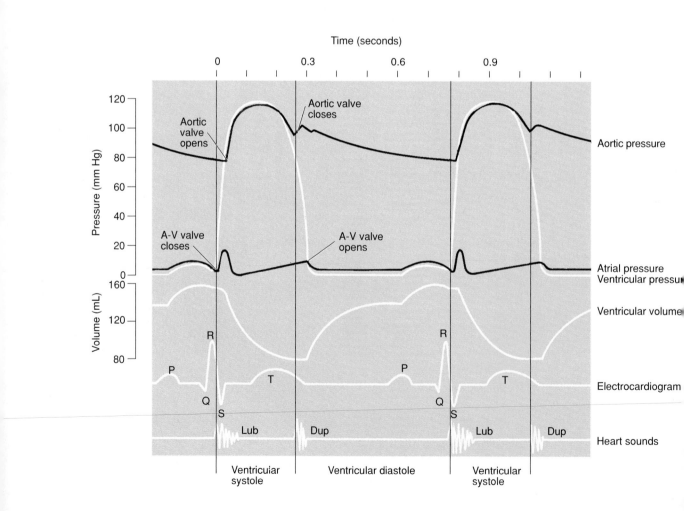

Graph of Cardiac Cycle, Left Ventricle Changes
Figure 13.11

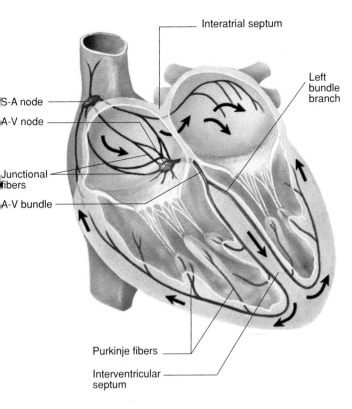

- Interatrial septum
- Left bundle branch
- S-A node
- A-V node
- Junctional fibers
- A-V bundle
- Purkinje fibers
- Interventricular septum

Cardiac Conduction System
Figure 13.12

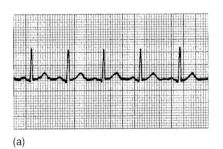

(a)

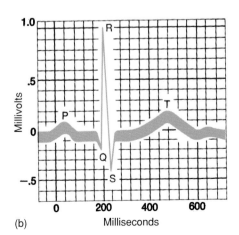

(b)

Normal ECG Pattern
Figure 13.15a,b

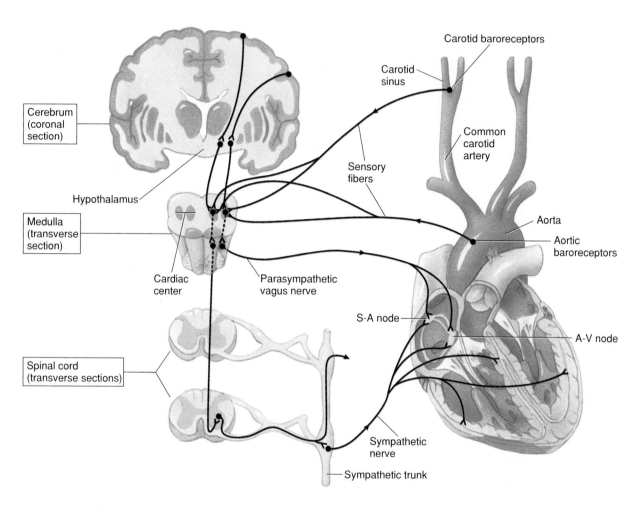

Cerebrum (coronal section)

Hypothalamus

Medulla (transverse section)

Cardiac center

Spinal cord (transverse sections)

Carotid baroreceptors

Carotid sinus

Common carotid artery

Sensory fibers

Aorta

Aortic baroreceptors

Parasympathetic vagus nerve

S-A node

A-V node

Sympathetic nerve

Sympathetic trunk

Autonomic Nerve Impulses to the Heart
Figure 13.17

Artery

Vein

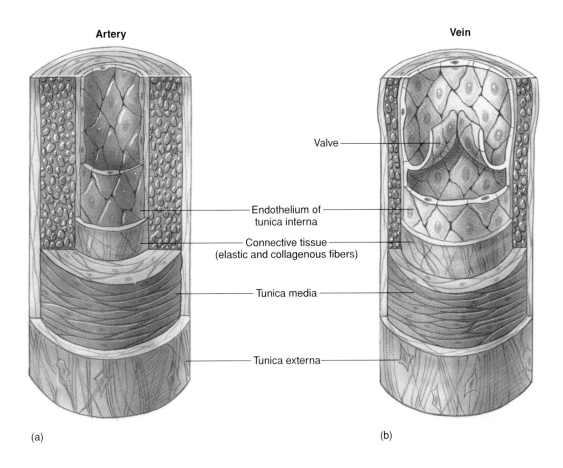

Valve

Endothelium of
tunica interna

Connective tissue
(elastic and collagenous fibers)

Tunica media

Tunica externa

(a)

(b)

Vessel Wall, Artery and Vein
Figure 13.18

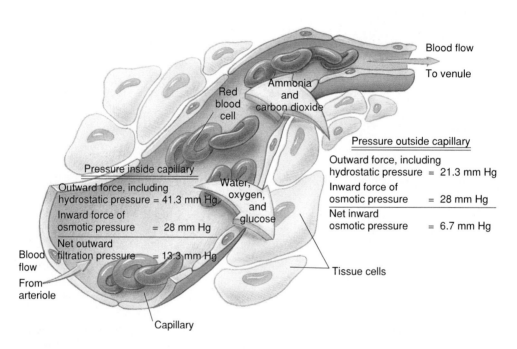

Blood flow
To venule

Ammonia
and
carbon dioxide

Red
blood
cell

Pressure outside capillary

Outward force, including
hydrostatic pressure = 21.3 mm Hg

Inward force of
osmotic pressure = 28 mm Hg

Net inward
osmotic pressure = 6.7 mm Hg

Pressure inside capillary

Outward force, including
hydrostatic pressure = 41.3 mm Hg

Inward force of
osmotic pressure = 28 mm Hg

Net outward
filtration pressure = 13.3 mm Hg

Water,
oxygen,
and
glucose

Blood
flow

From
arteriole

Tissue cells

Capillary

Capillary Exchanges
Figure 13.22

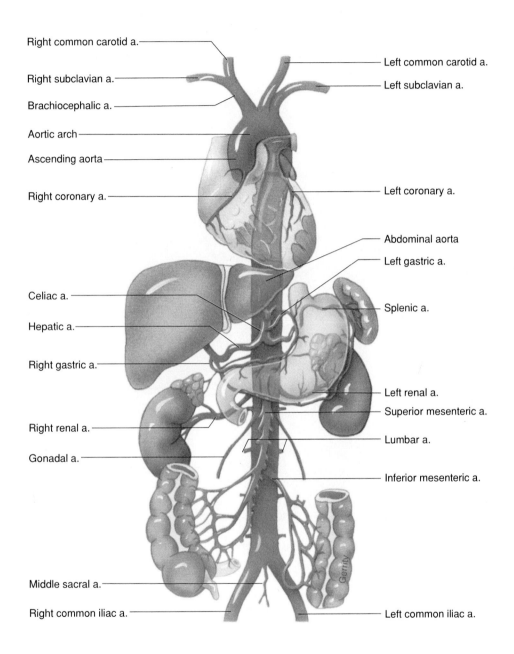

Right common carotid a.

Left common carotid a.

Right subclavian a.

Left subclavian a.

Brachiocephalic a.

Aortic arch

Ascending aorta

Right coronary a.

Left coronary a.

Abdominal aorta

Left gastric a.

Celiac a.

Splenic a.

Hepatic a.

Right gastric a.

Left renal a.

Superior mesenteric a.

Right renal a.

Lumbar a.

Gonadal a.

Inferior mesenteric a.

Middle sacral a.

Right common iliac a.

Left common iliac a.

Branches of the Aorta
Figure 13.26

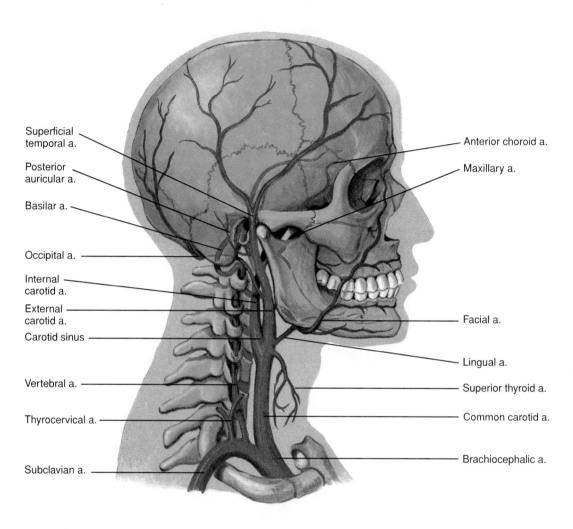

Superficial temporal a.

Posterior auricular a.

Basilar a.

Occipital a.

Internal carotid a.

External carotid a.

Carotid sinus

Vertebral a.

Thyrocervical a.

Subclavian a.

Anterior choroid a.

Maxillary a.

Facial a.

Lingual a.

Superior thyroid a.

Common carotid a.

Brachiocephalic a.

Arteries, Head and Neck
Figure 13.27

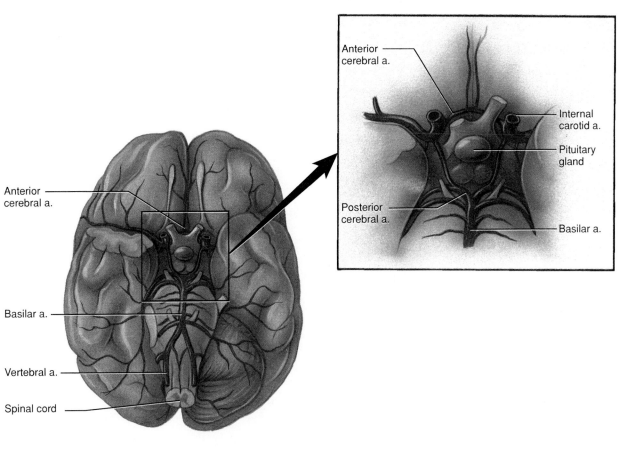

Anterior
cerebral a.

Internal
carotid a.

Pituitary
gland

Posterior
cerebral a.

Basilar a.

Anterior
cerebral a.

Basilar a.

Vertebral a.

Spinal cord

Cerebral Arteries
Figure 13.28

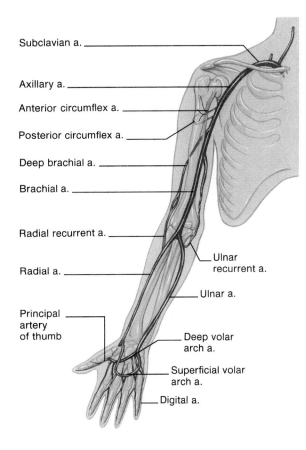

Subclavian a.

Axillary a.

Anterior circumflex a.

Posterior circumflex a.

Deep brachial a.

Brachial a.

Radial recurrent a.

Ulnar
recurrent a.

Radial a.

Ulnar a.

Principal
artery
of thumb

Deep volar
arch a.

Superficial volar
arch a.

Digital a.

Arteries, Shoulder and Upper Limb
Figure 13.29

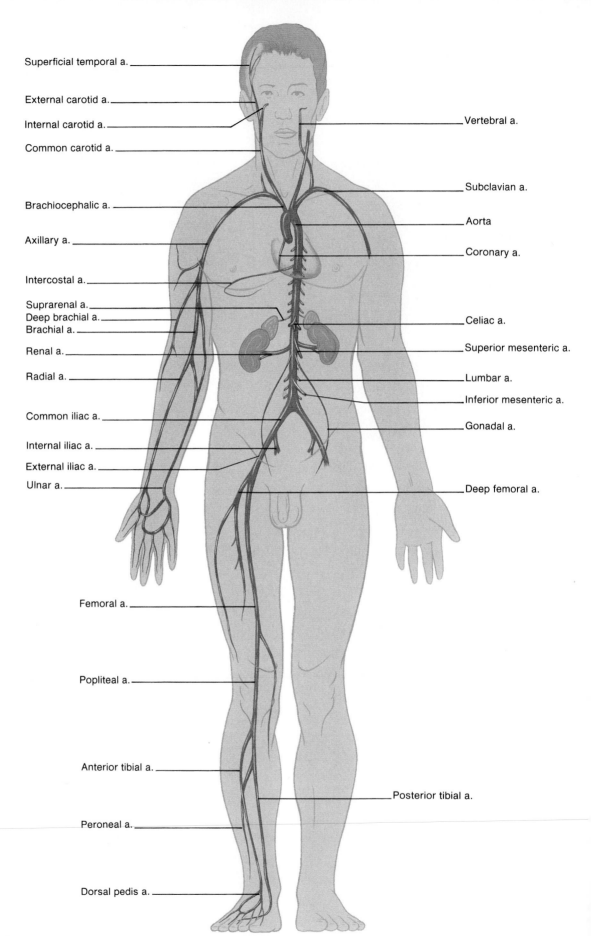

Superficial temporal a.

External carotid a.

Internal carotid a.

Common carotid a.

Brachiocephalic a.

Axillary a.

Intercostal a.

Suprarenal a.
Deep brachial a.
Brachial a.

Renal a.

Radial a.

Common iliac a.

Internal iliac a.

External iliac a.

Ulnar a.

Femoral a.

Popliteal a.

Anterior tibial a.

Peroneal a.

Dorsal pedis a.

Vertebral a.

Subclavian a.

Aorta

Coronary a.

Celiac a.

Superior mesenteric a.

Lumbar a.

Inferior mesenteric a.

Gonadal a.

Deep femoral a.

Posterior tibial a.

Major Arteries
Figure 13.31

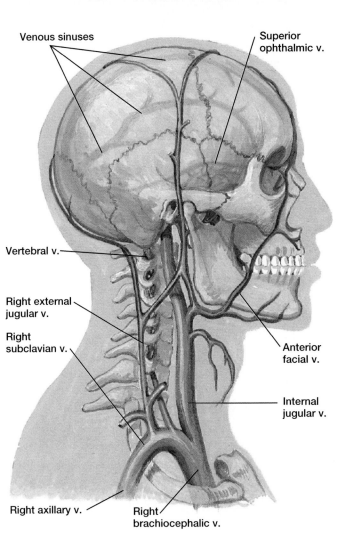

Venous sinuses

Superior
ophthalmic v.

Vertebral v.

Right external
jugular v.

Right
subclavian v.

Anterior
facial v.

Internal
jugular v.

Right axillary v.

Right
brachiocephalic v.

Veins, Head and Neck
Figure 13.32

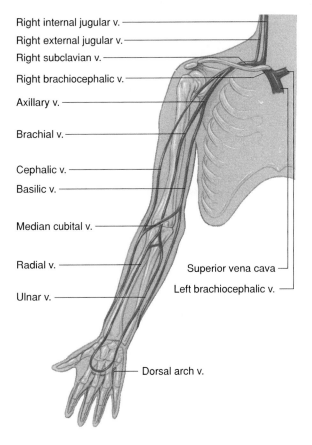

Right internal jugular v.

Right external jugular v.

Right subclavian v.

Right brachiocephalic v.

Axillary v.

Brachial v.

Cephalic v.

Basilic v.

Median cubital v.

Radial v.

Superior vena cava

Left brachiocephalic v.

Ulnar v.

Dorsal arch v.

Veins, Shoulder and Upper Limb
Figure 13.33

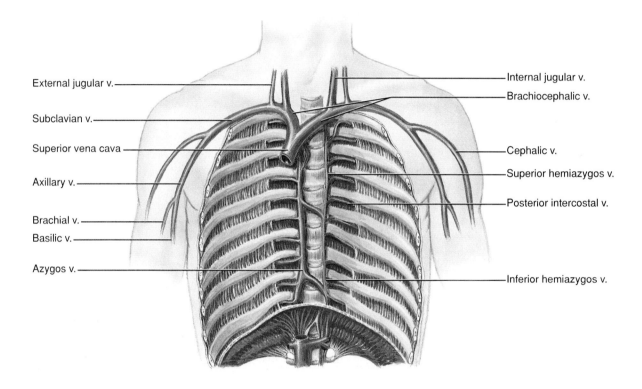

External jugular v.

Subclavian v.

Superior vena cava

Axillary v.

Brachial v.

Basilic v.

Azygos v.

Internal jugular v.

Brachiocephalic v.

Cephalic v.

Superior hemiazygos v.

Posterior intercostal v.

Inferior hemiazygos v.

Thoracic Veins
Figure 13.34

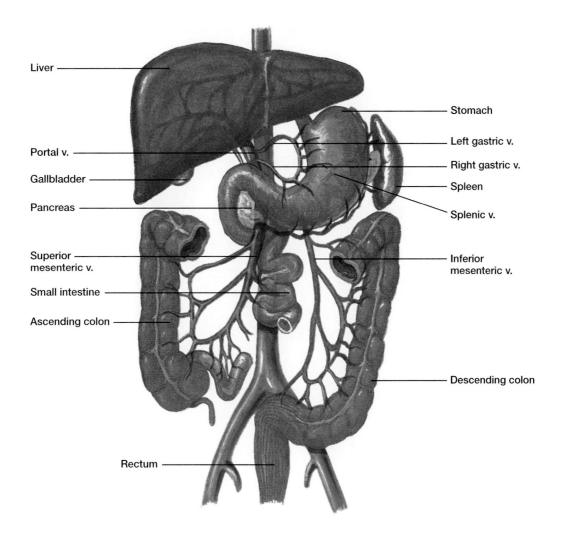

Liver

Stomach

Portal v.

Left gastric v.

Gallbladder

Right gastric v.

Pancreas

Spleen

Splenic v.

Superior
mesenteric v.

Inferior
mesenteric v.

Small intestine

Ascending colon

Descending colon

Rectum

Abdominal Veins
Figure 13.35

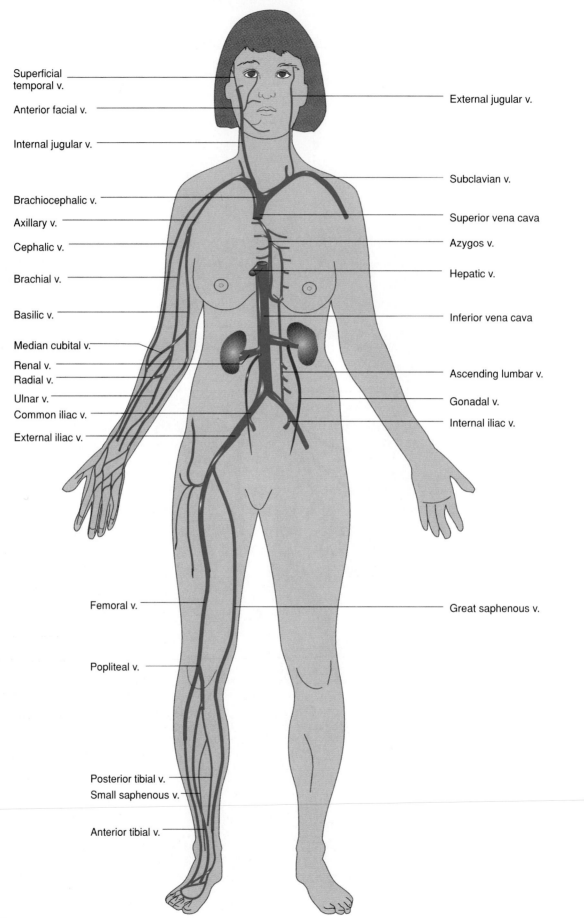

Superficial temporal v.

Anterior facial v.

Internal jugular v.

Brachiocephalic v.

Axillary v.

Cephalic v.

Brachial v.

Basilic v.

Median cubital v.

Renal v.

Radial v.

Ulnar v.

Common iliac v.

External iliac v.

External jugular v.

Subclavian v.

Superior vena cava

Azygos v.

Hepatic v.

Inferior vena cava

Ascending lumbar v.

Gonadal v.

Internal iliac v.

Femoral v.

Great saphenous v.

Popliteal v.

Posterior tibial v.

Small saphenous v.

Anterior tibial v.

Major Veins
Figure 13.36

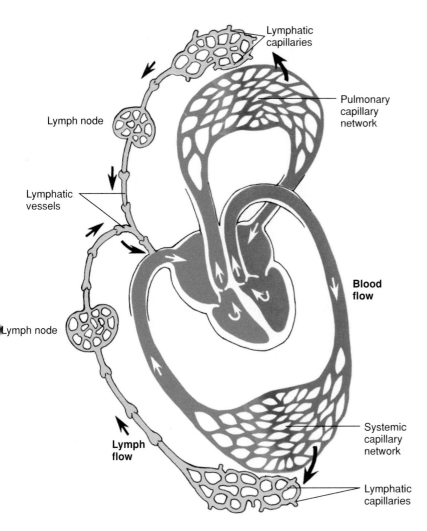

Lymph Transport
Figure 14.1

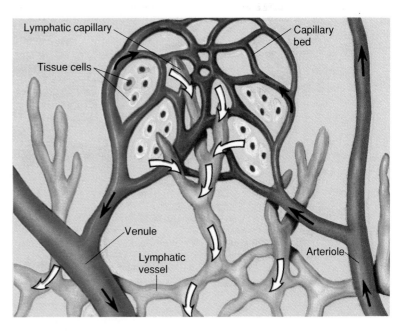

Lymph Capillaries
Figure 14.2

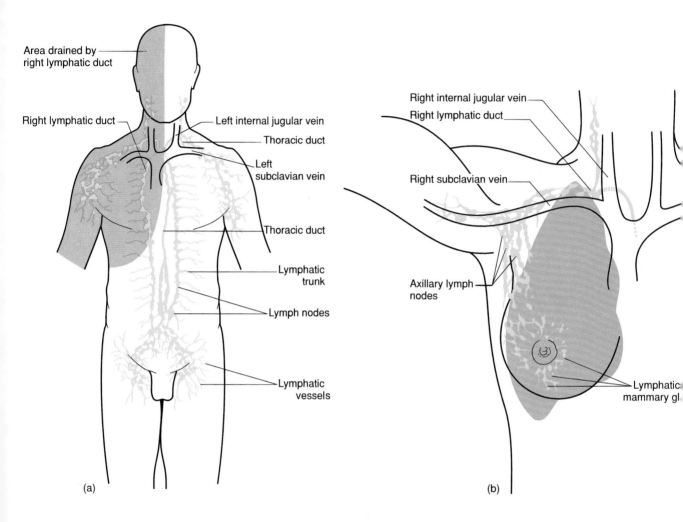

Lymph Drainage
Figure 14.4

Area drained by right lymphatic duct

Right lymphatic duct

Left internal jugular vein

Thoracic duct

Left subclavian vein

Thoracic duct

Lymphatic trunk

Lymph nodes

Lymphatic vessels

(a)

Right internal jugular vein

Right lymphatic duct

Right subclavian vein

Axillary lymph nodes

Lymphatic mammary gl

(b)

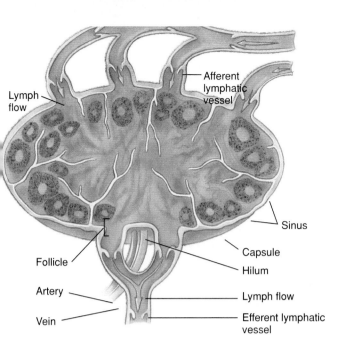

Lymph
flow

Afferent
lymphatic
vessel

Sinus

Capsule

Hilum

Follicle

Artery

Vein

Lymph flow

Efferent lymphatic
vessel

Lymph Node
Figure 14.6

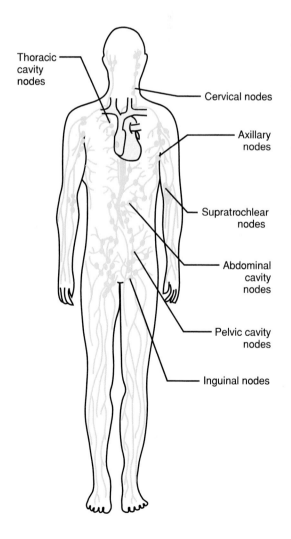

Thoracic
cavity
nodes

Cervical nodes

Axillary
nodes

Supratrochlear
nodes

Abdominal
cavity
nodes

Pelvic cavity
nodes

Inguinal nodes

Lymph Node Locations
Figure 14.8

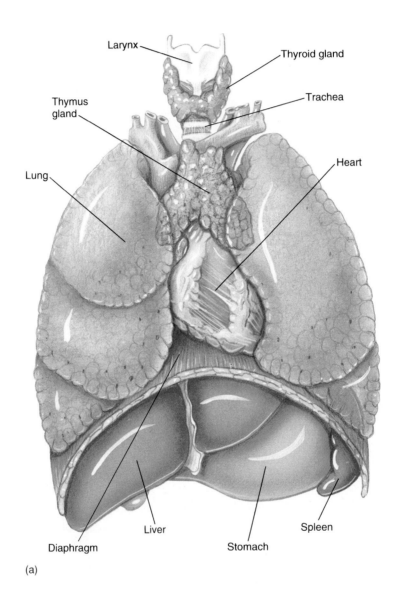

Larynx

Thyroid gland

Thymus
gland

Trachea

Lung

Heart

Liver

Diaphragm

Stomach

Spleen

(a)

The Thymus
Figure 14.9a

B cells

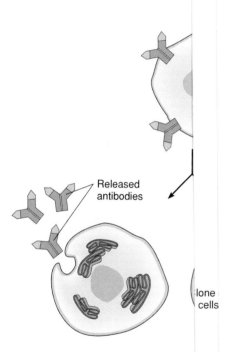

Released
antibodies

clone
cells

Plasma cell
(antibody-secreting cell)

Plasma and Memory Cells
Figure 14.16

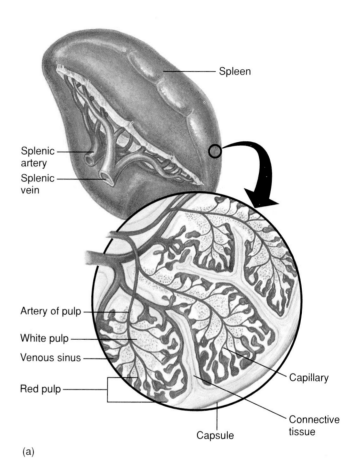

Spleen

Splenic
artery

Splenic
vein

Artery of pulp

White pulp

Venous sinus

Red pulp

Capsule

Capillary

Connective
tissue

(a)

The Spleen
Figure 14.10a

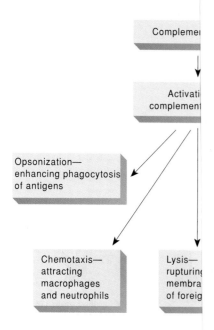

Complemer

↓

Activati
complement

Opsonization—
enhancing phagocytosis
of antigens

Chemotaxis—
attracting
macrophages
and neutrophils

Lysis—
rupturing
membra
of foreig

The Complement System
Figure 14.14

Antigen
receptor

Antigen

Displaye

(1)

B cell and
Figure 14.

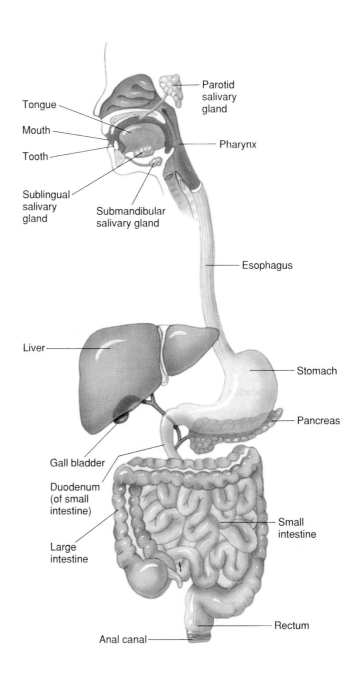

Tongue

Mouth

Tooth

Sublingual
salivary
gland

Submandibular
salivary gland

Parotid
salivary
gland

Pharynx

Esophagus

Liver

Stomach

Pancreas

Gall bladder

Duodenum
(of small
intestine)

Large
intestine

Small
intestine

Rectum

Anal canal

Digestive System
Figure 15.1

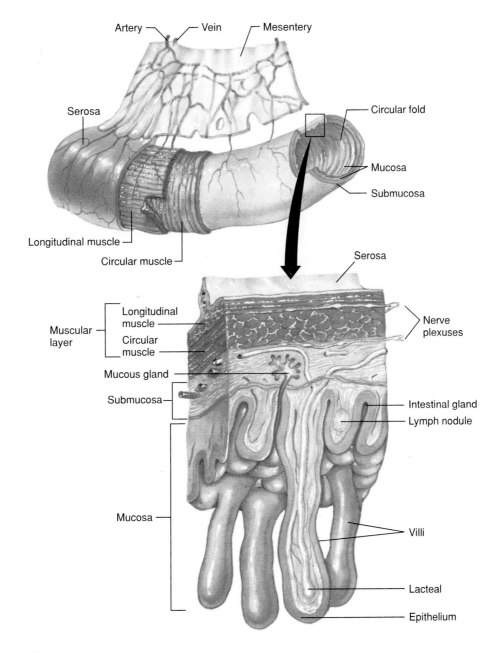

Wall of the Alimentary Canal
Figure 15.3

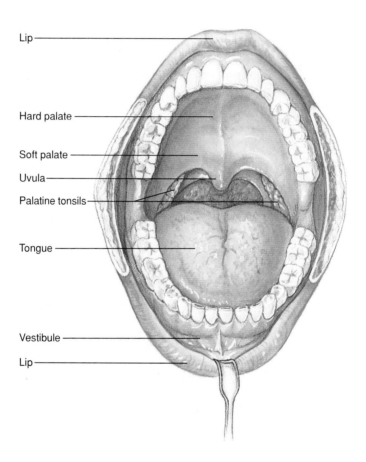

Lip

Hard palate

Soft palate

Uvula

Palatine tonsils

Tongue

Vestibule

Lip

The Mouth
Figure 15.5

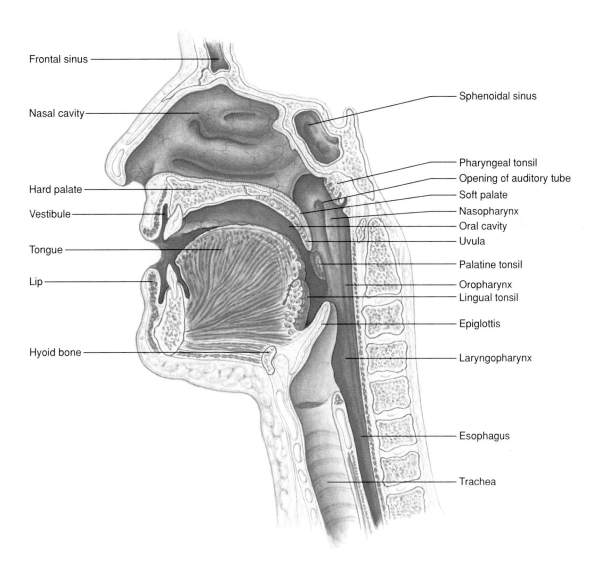

Frontal sinus

Nasal cavity

Hard palate

Vestibule

Tongue

Lip

Hyoid bone

Sphenoidal sinus

Pharyngeal tonsil

Opening of auditory tube

Soft palate

Nasopharynx

Oral cavity

Uvula

Palatine tonsil

Oropharynx

Lingual tonsil

Epiglottis

Laryngopharynx

Esophagus

Trachea

Mouth, Nasal Cavity, and Pharynx
Figure 15.6

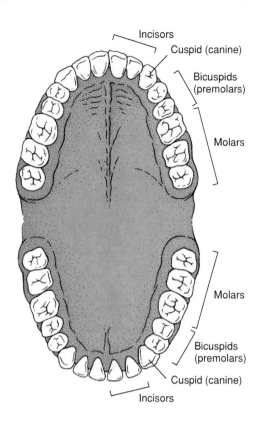

Incisors
Cuspid (canine)
Bicuspids
(premolars)

Molars

Molars

Bicuspids
(premolars)
Cuspid (canine)
Incisors

The Teeth
Figure 15.8

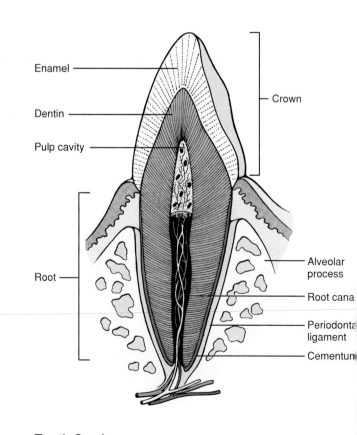

Enamel

Dentin

Pulp cavity

Crown

Root

Alveolar
process

Root cana

Periodonta
ligament

Cementun

Tooth Section
Figure 15.9

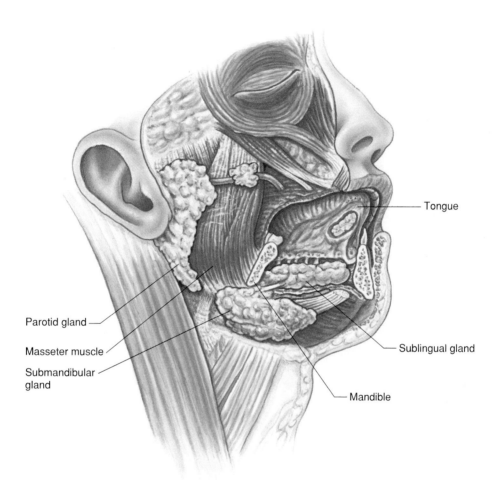

Tongue

Parotid gland

Masseter muscle

Submandibular
gland

Sublingual gland

Mandible

The Salivary Glands
Figure 15.10

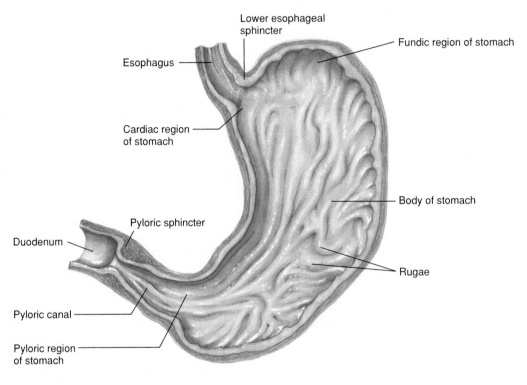

Lower esophageal
sphincter

Fundic region of stomach

Esophagus

Cardiac region
of stomach

Body of stomach

Pyloric sphincter

Duodenum

Rugae

Pyloric canal

Pyloric region
of stomach

The Stomach
Figure 15.11

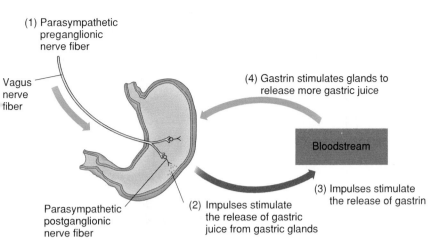

(1) Parasympathetic
preganglionic
nerve fiber

Vagus
nerve
fiber

(4) Gastrin stimulates glands to
release more gastric juice

Bloodstream

(3) Impulses stimulate
the release of gastrin

Parasympathetic
postganglionic
nerve fiber

(2) Impulses stimulate
the release of gastric
juice from gastric glands

Gastric Juice Secretion
Figure 15.14

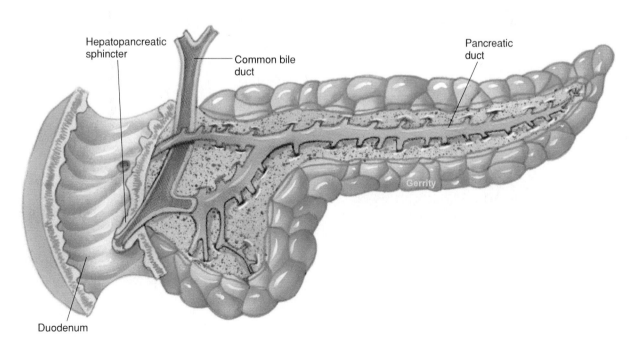

Hepatopancreatic
sphincter

Common bile
duct

Pancreatic
duct

Gerrity

Duodenum

The Pancreas
Figure 15.15

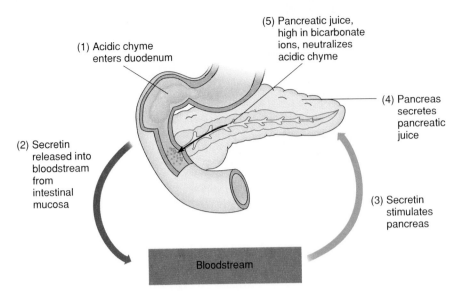

(1) Acidic chyme enters duodenum

(5) Pancreatic juice, high in bicarbonate ions, neutralizes acidic chyme

(4) Pancreas secretes pancreatic juice

(2) Secretin released into bloodstream from intestinal mucosa

(3) Secretin stimulates pancreas

Bloodstream

Pancreatic Juice Secretion
Figure 15.16

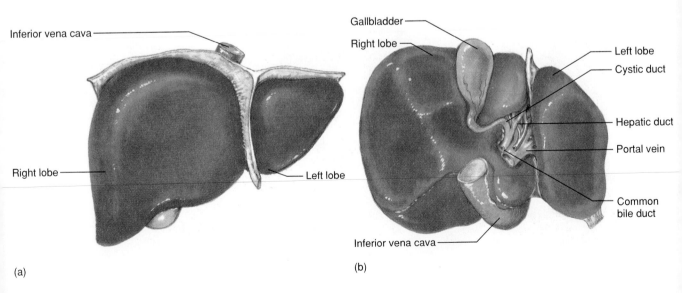

Inferior vena cava

Right lobe

Left lobe

(a)

Gallbladder

Right lobe

Left lobe

Cystic duct

Hepatic duct

Portal vein

Common bile duct

Inferior vena cava

(b)

The Liver
Figure 15.17

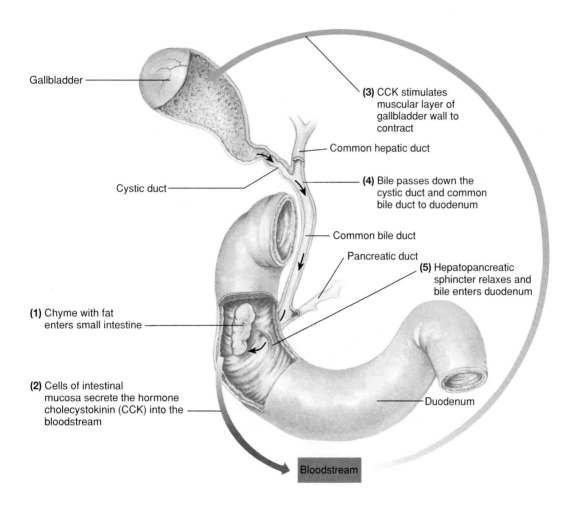

Gallbladder

(3) CCK stimulates muscular layer of gallbladder wall to contract

Common hepatic duct

Cystic duct

(4) Bile passes down the cystic duct and common bile duct to duodenum

Common bile duct

Pancreatic duct

(5) Hepatopancreatic sphincter relaxes and bile enters duodenum

(1) Chyme with fat enters small intestine

(2) Cells of intestinal mucosa secrete the hormone cholecystokinin (CCK) into the bloodstream

Duodenum

Bloodstream

Bile Release
Figure 15.21

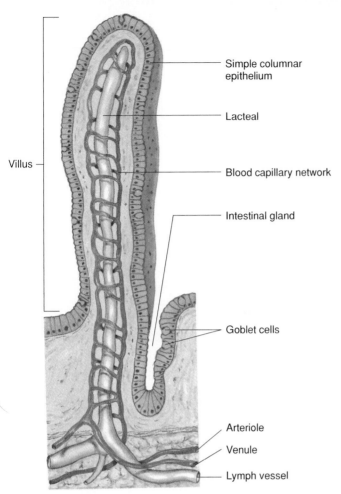

Intestinal Villus
Figure 15.25

Villus

Simple columnar epithelium

Lacteal

Blood capillary network

Intestinal gland

Goblet cells

Arteriole

Venule

Lymph vessel

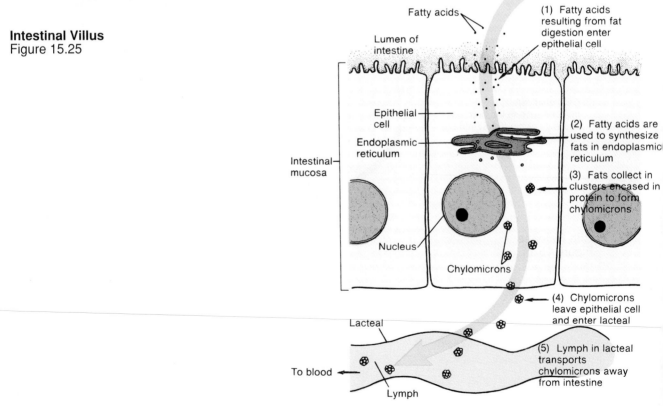

Fatty acids

Lumen of intestine

(1) Fatty acids resulting from fat digestion enter epithelial cell

Epithelial cell

Endoplasmic reticulum

Intestinal mucosa

(2) Fatty acids are used to synthesize fats in endoplasmic reticulum

(3) Fats collect in clusters encased in protein to form chylomicrons

Nucleus

Chylomicrons

(4) Chylomicrons leave epithelial cell and enter lacteal

Lacteal

To blood

(5) Lymph in lacteal transports chylomicrons away from intestine

Lymph

Fatty Acid Absorption
Figure 15.27

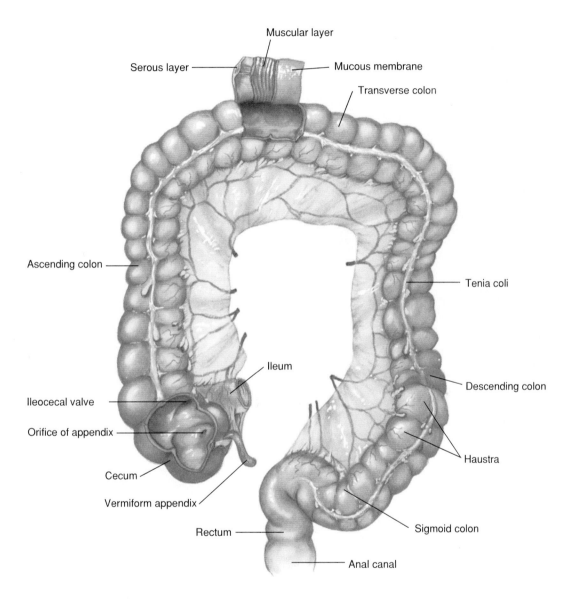

Muscular layer

Serous layer

Mucous membrane

Transverse colon

Ascending colon

Tenia coli

Ileum

Descending colon

Ileocecal valve

Orifice of appendix

Haustra

Cecum

Vermiform appendix

Rectum

Sigmoid colon

Anal canal

The Large Intestine
Figure 15.28

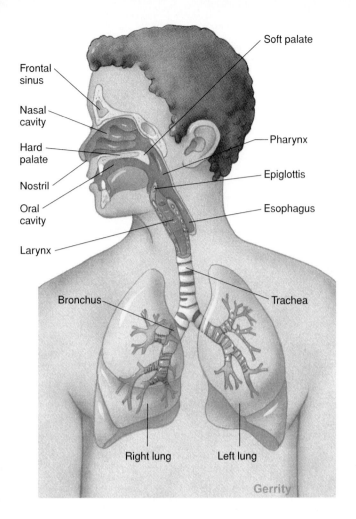

Frontal
sinus

Nasal
cavity

Hard
palate

Nostril

Oral
cavity

Larynx

Soft palate

Pharynx

Epiglottis

Esophagus

Bronchus

Trachea

Right lung

Left lung

Gerrity

Respiratory System
Figure 16.1

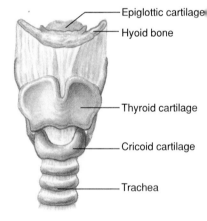

Epiglottic cartilage

Hyoid bone

Thyroid cartilage

Cricoid cartilage

Trachea

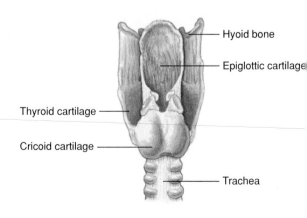

Hyoid bone

Epiglottic cartilage

Thyroid cartilage

Cricoid cartilage

Trachea

Larynx
Figure 16.4

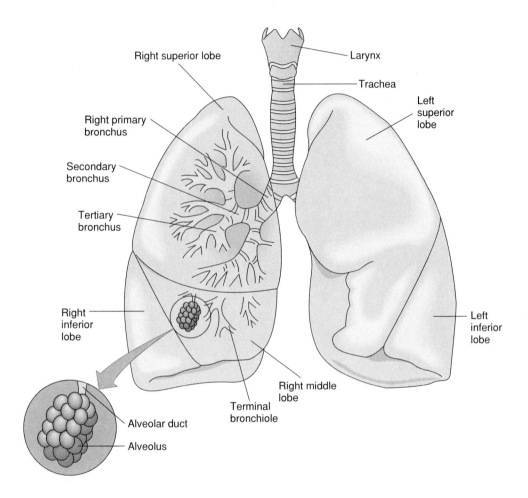

Right superior lobe

Right primary bronchus

Secondary bronchus

Tertiary bronchus

Right inferior lobe

Larynx

Trachea

Left superior lobe

Left inferior lobe

Right middle lobe

Terminal bronchiole

Alveolar duct

Alveolus

Bronchial Tree
Figure 16.7

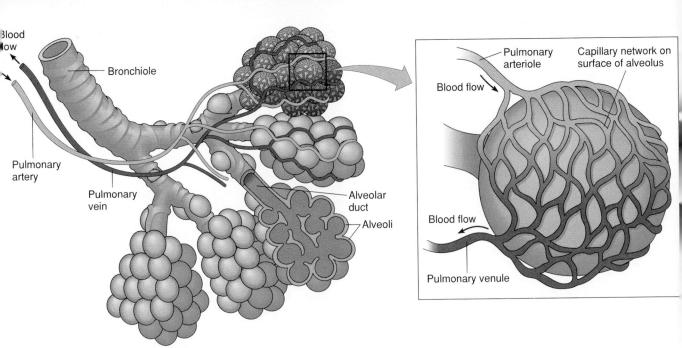

Respiratory Tubes and Alveoli
Figure 16.8

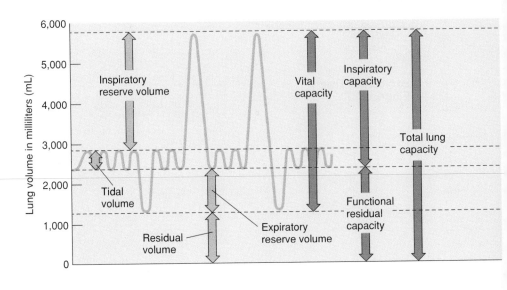

Lung Volumes
Figure 16.14

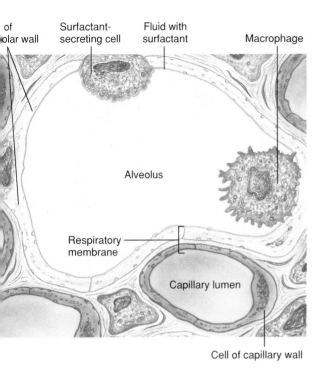

of
olar wall

Surfactant-
secreting cell

Fluid with
surfactant

Macrophage

Alveolus

Respiratory
membrane

Capillary lumen

Cell of capillary wall

spiratory Membrane
ure 16.18

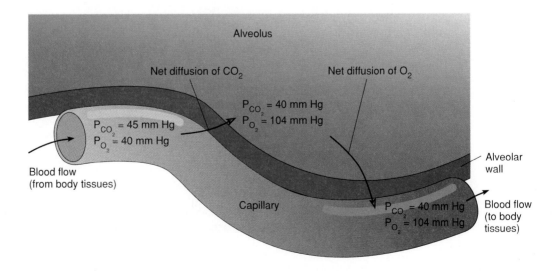

Alveolus

Net diffusion of CO_2

Net diffusion of O_2

P_{CO_2} = 40 mm Hg
P_{O_2} = 104 mm Hg

P_{CO_2} = 45 mm Hg
P_{O_2} = 40 mm Hg

Blood flow
(from body tissues)

Capillary

Alveolar
wall

P_{CO_2} = 40 mm Hg
P_{O_2} = 104 mm Hg

Blood flow
(to body
tissues)

Alveolar Gas Exchange
Figure 16.19

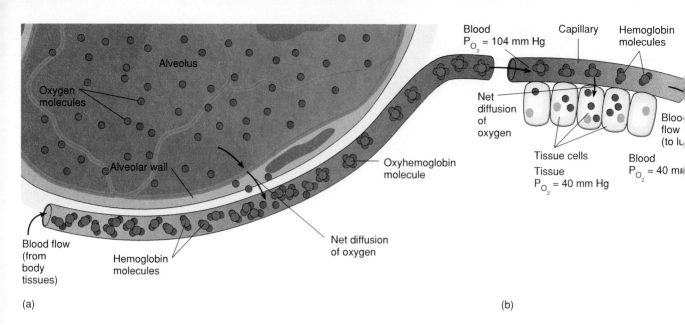

Oxyhemoglobin
Figure 16.20

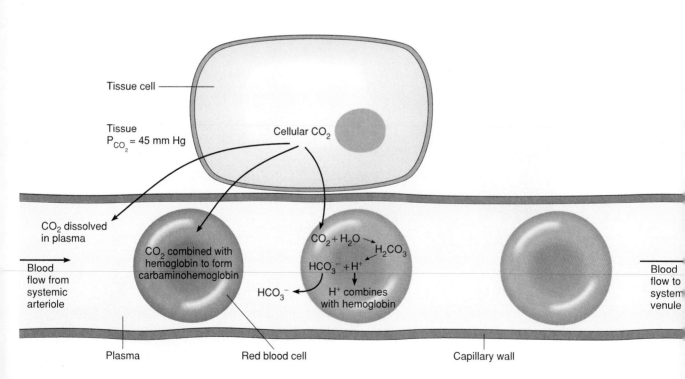

Carbon Dioxide Transport I
Figure 16.21

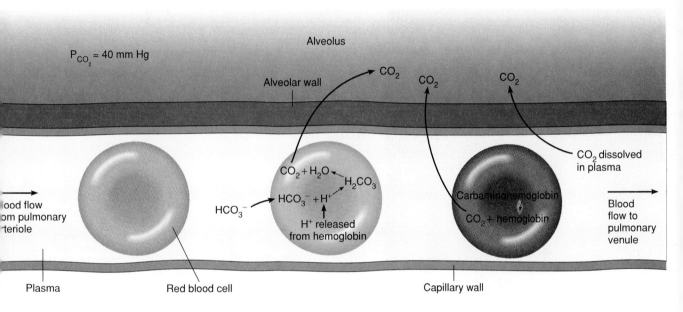

$P_{CO_2} = 40$ mm Hg

Alveolus

Alveolar wall

CO$_2$

CO$_2$

CO$_2$

Blood flow from pulmonary arteriole

CO$_2$ + H$_2$O $\leftarrow$ H$_2$CO$_3$

HCO$_3^-$ $\rightarrow$ HCO$_3^-$ + H$^+$

H$^+$ released from hemoglobin

Carbaminohemoglobin

CO$_2$ + hemoglobin

CO$_2$ dissolved in plasma

Blood flow to pulmonary venule

Plasma

Red blood cell

Capillary wall

rbon Dioxide Transport II

gure 16.22

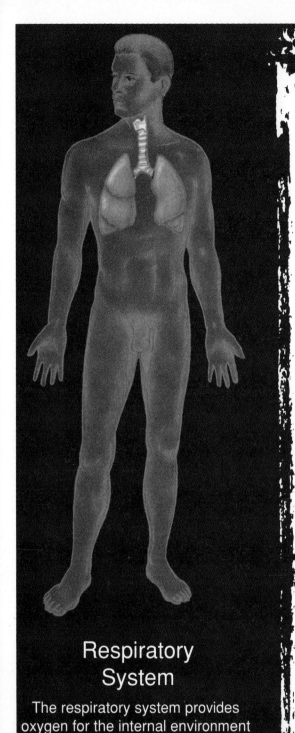

Respiratory
System

The respiratory system provides oxygen for the internal environment and excretes carbon dioxide.

Integumentary System

Stimulation of skin receptors may alter respiratory rate.

Cardiovascular System

As the heart pumps blood through the lungs, the lungs oxygenate the blood and excrete carbon dioxide.

Skeletal System

Bones provide attachments for muscles involved in breathing.

Lymphatic System

Cells of the immune system patrol the lungs and defend against infection.

Muscular System

The respiratory system eliminates carbon dioxide produced by exercising muscles.

Digestive System

The digestive system and respiratory system share openings to the outside.

Nervous System

The brain controls the respiratory system. The respiratory system helps control pH of the internal environment.

Urinary System

The kidneys and the respiratory system work together to maintain blood pH. The kidneys compensate for water lost through breathing.

Endocrine System

Hormonelike substances control the production of red blood cells that transport oxygen and carbon dioxide.

Reproductive System

Respiration increases during sexual activity. Fetal "respiration" begins before birth.

Respiratory System
ORGANIZATION Chapter 16

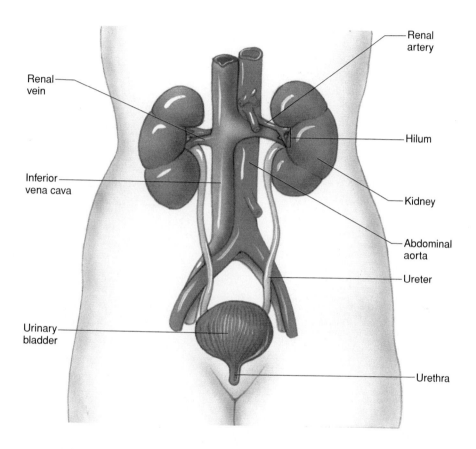

Renal artery

Renal vein

Hilum

Inferior vena cava

Kidney

Abdominal aorta

Ureter

Urinary bladder

Urethra

Urinary System
Figure 17.1

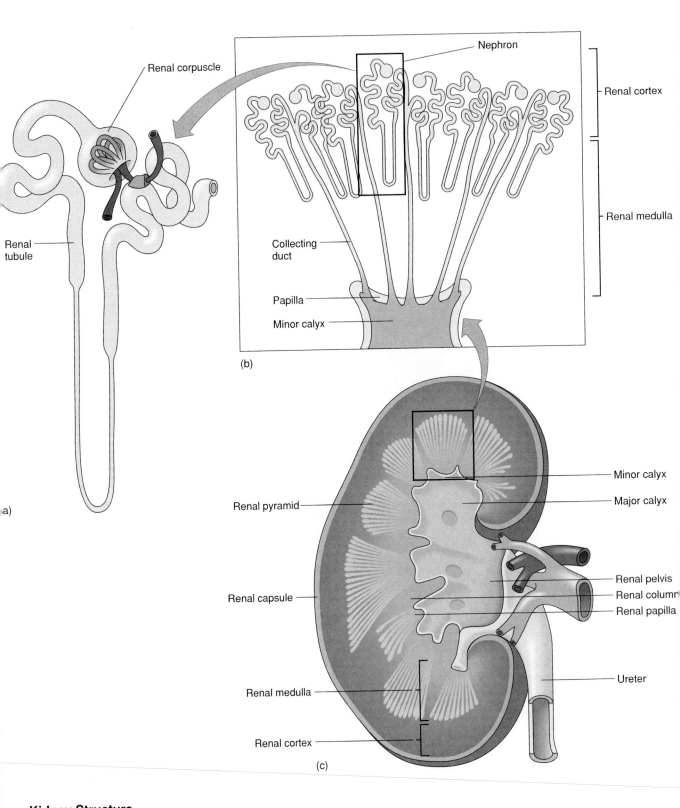

Renal corpuscle

Renal tubule

(a)

Nephron

Renal cortex

Renal medulla

Collecting duct

Papilla

Minor calyx

(b)

Renal pyramid

Renal capsule

Renal medulla

Renal cortex

(c)

Minor calyx

Major calyx

Renal pelvis

Renal column

Renal papilla

Ureter

Kidney Structure
Figure 17.2

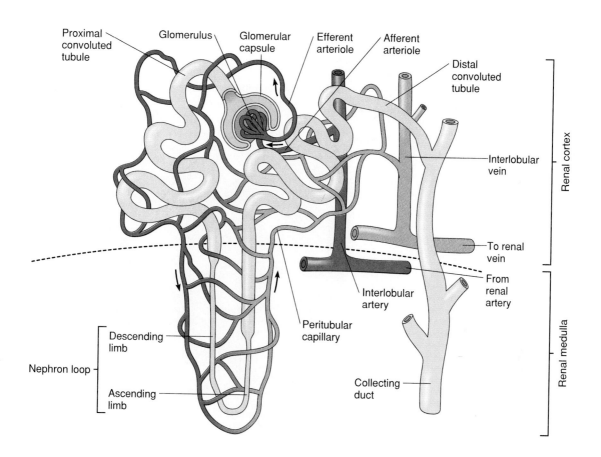

Proximal convoluted tubule

Glomerulus

Glomerular capsule

Efferent arteriole

Afferent arteriole

Distal convoluted tubule

Renal cortex

Interlobular vein

To renal vein

From renal artery

Interlobular artery

Peritubular capillary

Descending limb

Nephron loop

Ascending limb

Renal medulla

Collecting duct

Nephron Structure
Figure 17.6

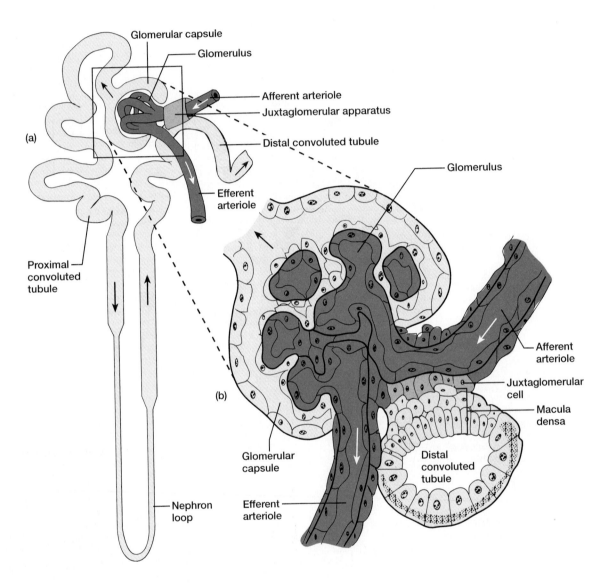

(a)

Glomerular capsule

Glomerulus

Afferent arteriole

Juxtaglomerular apparatus

Distal convoluted tubule

Efferent arteriole

Proximal convoluted tubule

Nephron loop

(b)

Glomerulus

Afferent arteriole

Juxtaglomerular cell

Macula densa

Distal convoluted tubule

Glomerular capsule

Efferent arteriole

Juxtaglomerular Apparatus
Figure 17.7

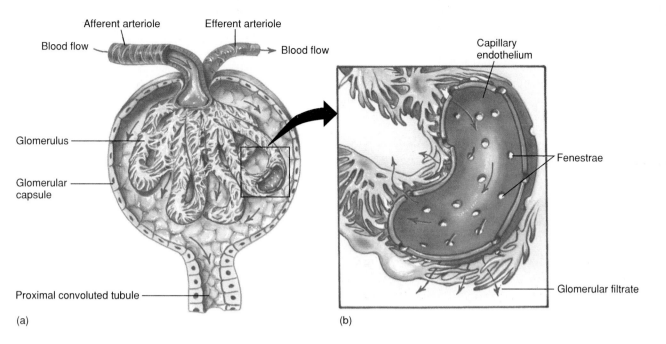

Glomerular Filtration
Figure 17.8

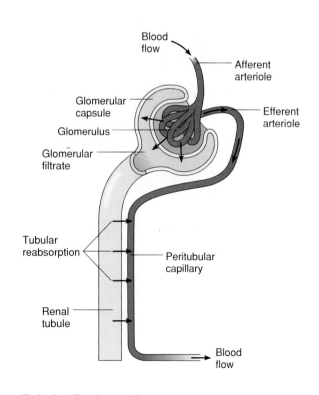

Tubular Reabsorption
Figure 17.10

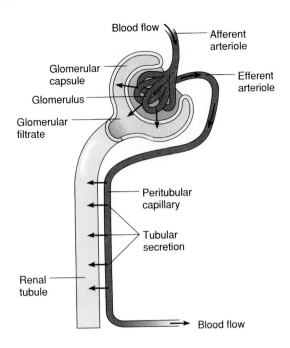

Blood flow
Afferent arteriole
Glomerular capsule
Efferent arteriole
Glomerulus
Glomerular filtrate
Peritubular capillary
Tubular secretion
Renal tubule
Blood flow

Tubular Secretion
Figure 17.12

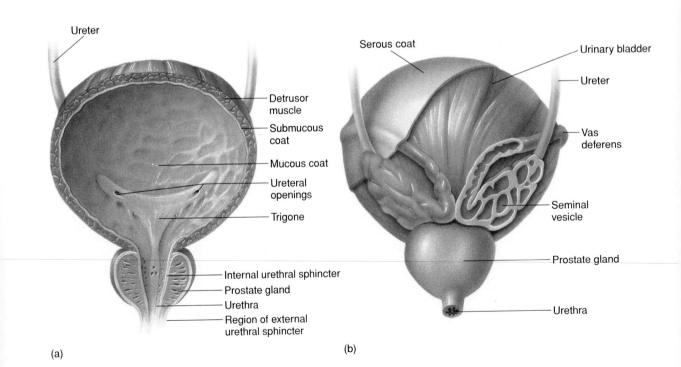

Ureter
Detrusor muscle
Submucous coat
Mucous coat
Ureteral openings
Trigone
Internal urethral sphincter
Prostate gland
Urethra
Region of external urethral sphincter

(a)

Serous coat
Urinary bladder
Ureter
Vas deferens
Seminal vesicle
Prostate gland
Urethra

(b)

Urinary Bladder
Figure 17.15

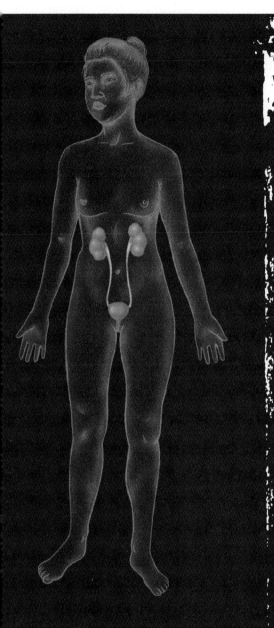

Urinary System

The urinary system controls the composition of the internal environment.

Integumentary System

The urinary system compensates for water loss due to sweating. The kidneys and skin both play a role in vitamin D production.

Cardiovascular System

The urinary system controls blood volume. Blood volume and blood pressure play a role in determining water and solute excretion.

Skeletal System

The kidneys and bone work together to control plasma calcium levels.

Lymphatic System

The kidneys control extracellular fluid volume and composition (including lymph).

Muscular System

Muscle tissue controls urine elimination from the bladder.

Digestive System

The kidneys compensate for fluids lost by the digestive system.

Nervous System

The nervous system influences urine production and elimination.

Respiratory System

The kidneys and the lungs work together to control the pH of the internal environment.

Endocrine System

The endocrine system influences urine production.

Reproductive System

The urinary system in males shares common organs with the reproductive system. The kidneys compensate for fluids lost from the male and female reproductive systems.

Urinary System
ORGANIZATION Chapter 17

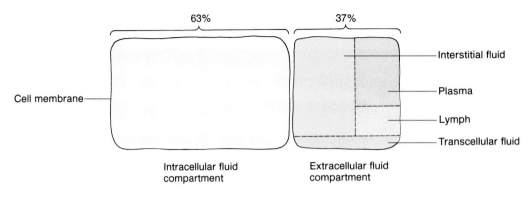

Total body water

63%

37%

Cell membrane

Interstitial fluid

Plasma

Lymph

Transcellular fluid

Intracellular fluid
compartment

Extracellular fluid
compartment

Fluid Compartments
Figure 18.1

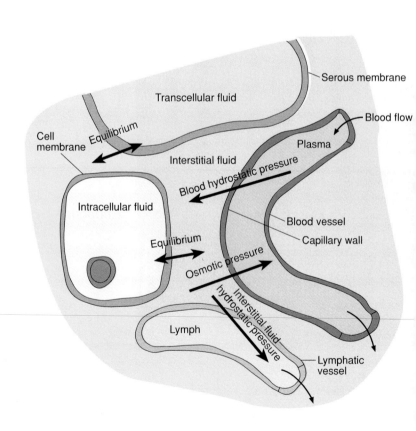

Serous membrane

Transcellular fluid

Blood flow

Cell
membrane

Equilibrium

Plasma

Interstitial fluid

Blood hydrostatic pressure

Intracellular fluid

Blood vessel

Capillary wall

Equilibrium

Osmotic pressure

Interstitial fluid
hydrostatic pressure

Lymph

Lymphatic
vessel

Fluid Movement
Figure 18.3

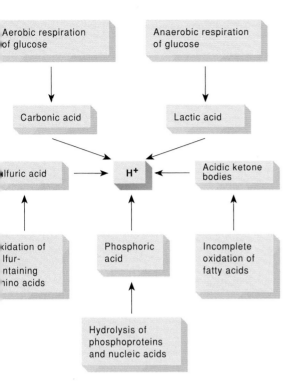

Aerobic respiration of glucose

Anaerobic respiration of glucose

Carbonic acid

Lactic acid

ılfuric acid → H^+ ← Acidic ketone bodies

xidation of lfur- ntaining nino acids

Phosphoric acid

Incomplete oxidation of fatty acids

Hydrolysis of phosphoproteins and nucleic acids

tabolic Sources of Hydrogen Ions
ure 18.7

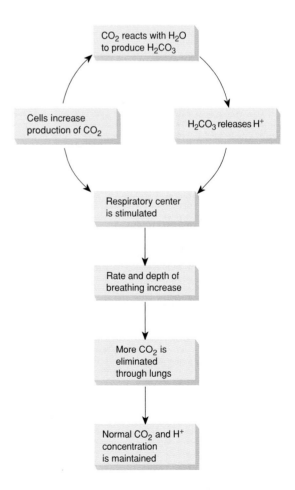

CO_2 reacts with H_2O to produce H_2CO_3

Cells increase production of CO_2

H_2CO_3 releases H^+

Respiratory center is stimulated

Rate and depth of breathing increase

More CO_2 is eliminated through lungs

Normal CO_2 and H^+ concentration is maintained

Increased Carbon Dioxide
Figure 18.8

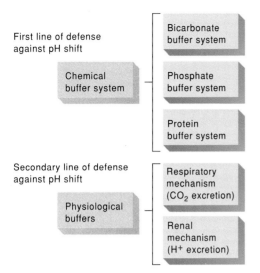

First line of defense
against pH shift

Chemical
buffer system

Bicarbonate
buffer system

Phosphate
buffer system

Protein
buffer system

Secondary line of defense
against pH shift

Physiological
buffers

Respiratory
mechanism
(CO_2 excretion)

Renal
mechanism
(H^+ excretion)

Buffers, Chemical and Physiological
Figure 18.9

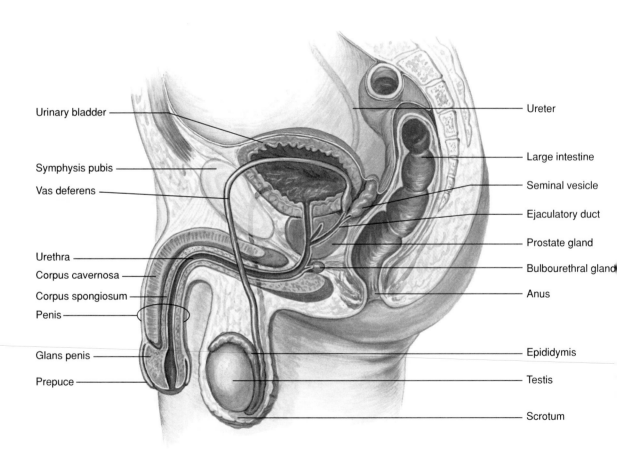

Urinary bladder

Symphysis pubis

Vas deferens

Urethra

Corpus cavernosa

Corpus spongiosum

Penis

Glans penis

Prepuce

Ureter

Large intestine

Seminal vesicle

Ejaculatory duct

Prostate gland

Bulbourethral gland

Anus

Epididymis

Testis

Scrotum

Male Reproductive System, Sagittal
Figure 19.1

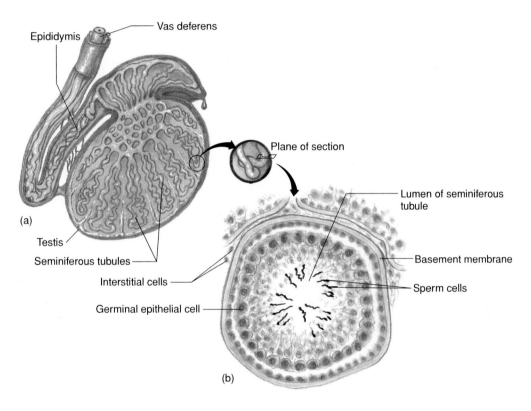

Testis and Seminiferous Tubule
Figure 19.2

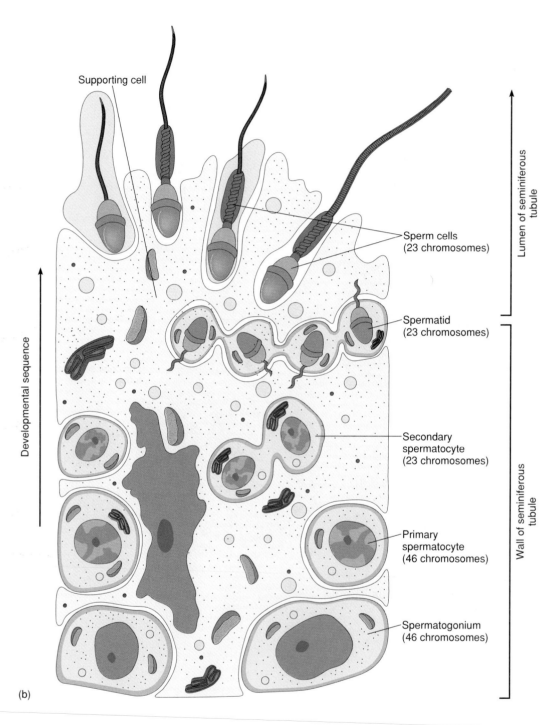

Supporting cell

Lumen of seminiferous tubule

Developmental sequence

Sperm cells
(23 chromosomes)

Spermatid
(23 chromosomes)

Secondary
spermatocyte
(23 chromosomes)

Wall of seminiferous tubule

Primary
spermatocyte
(46 chromosomes)

Spermatogonium
(46 chromosomes)

(b)

Spermatogonia
Figure 19.4b

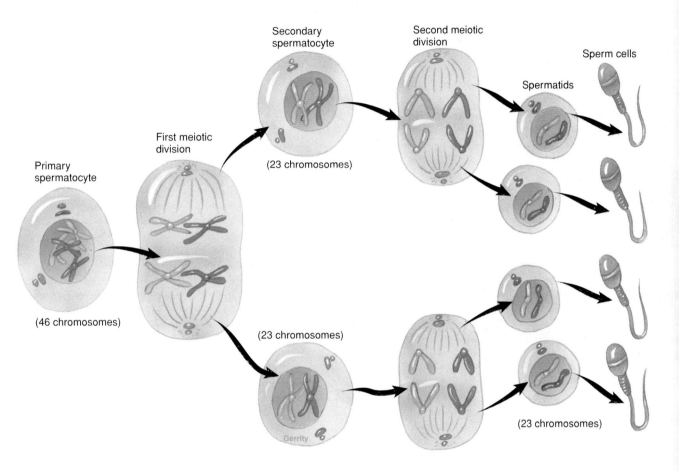

Secondary
spermatocyte

Second meiotic
division

Sperm cells

Spermatids

First meiotic
division

(23 chromosomes)

Primary
spermatocyte

(23 chromosomes)

(46 chromosomes)

Gerrity

(23 chromosomes)

Spermatogenesis
Figure 19.5

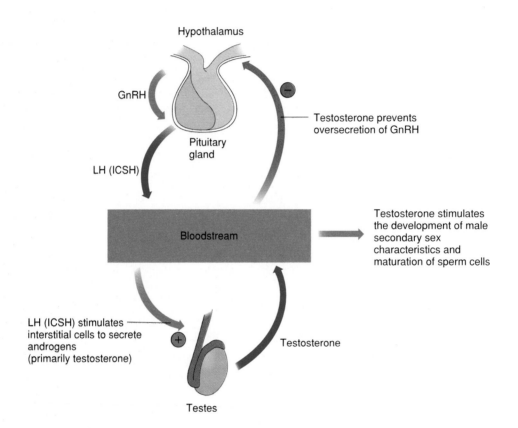

Male Sexual Development
Figure 19.6

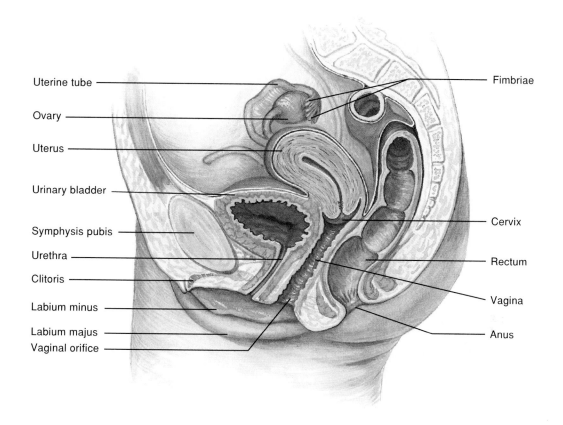

Uterine tube

Ovary

Uterus

Urinary bladder

Symphysis pubis

Urethra

Clitoris

Labium minus

Labium majus

Vaginal orifice

Fimbriae

Cervix

Rectum

Vagina

Anus

Female Reproductive System, Sagittal
Figure 19.7

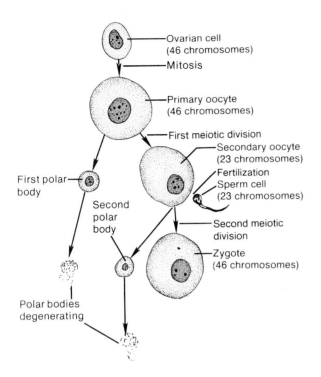

Oogenesis
Figure 19.8

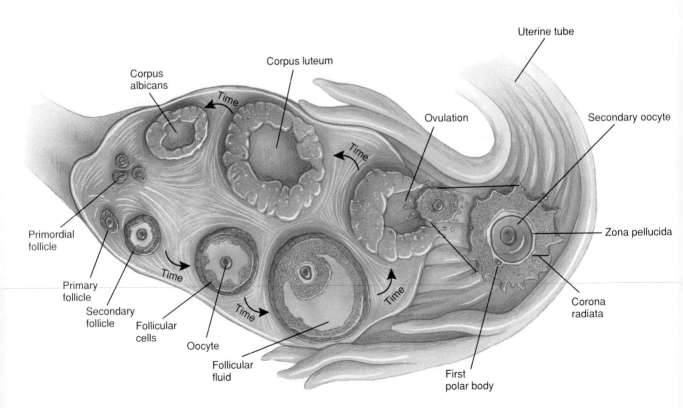

Ovarian Cycle
Figure 19.9

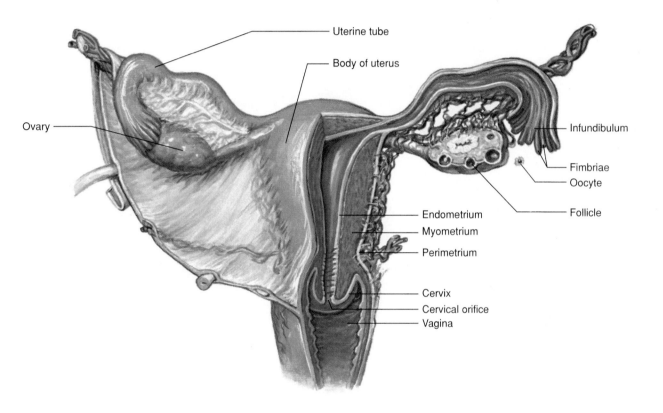

Uterine tube

Body of uterus

Ovary

Infundibulum

Fimbriae

Oocyte

Follicle

Endometrium

Myometrium

Perimetrium

Cervix

Cervical orifice

Vagina

Female Reproductive System, Frontal
Figure 19.11

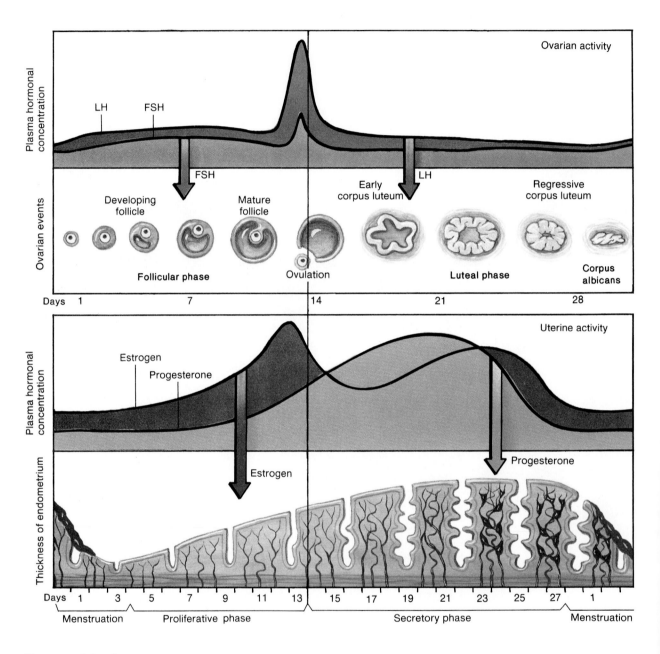

Menstrual Cycle
Figure 19.13

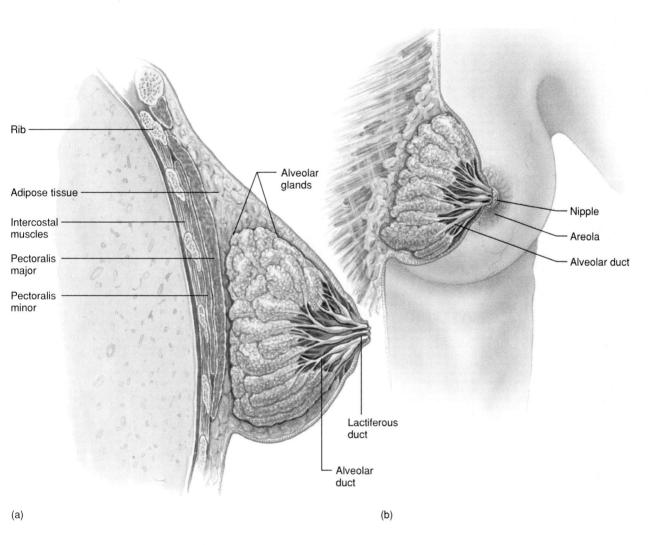

Rib

Adipose tissue

Intercostal
muscles

Pectoralis
major

Pectoralis
minor

Alveolar
glands

Lactiferous
duct

Alveolar
duct

Nipple

Areola

Alveolar duct

(a)

(b)

The Breast
Figure 19.14

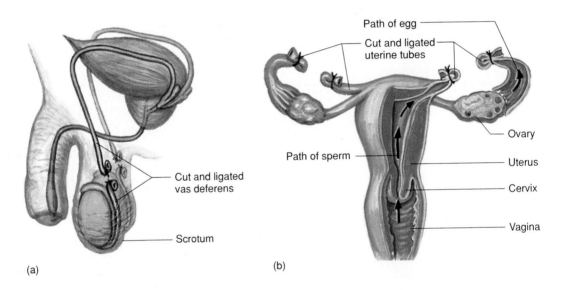

(a)

(b)

Cut and ligated
vas deferens

Scrotum

Path of egg

Cut and ligated
uterine tubes

Path of sperm

Ovary

Uterus

Cervix

Vagina

Vasectomy and Tubal Ligation
Figure 19.16

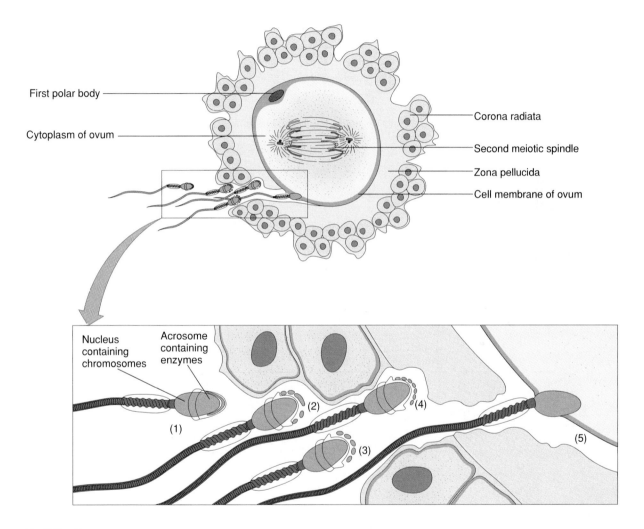

First polar body

Cytoplasm of ovum

Corona radiata

Second meiotic spindle

Zona pellucida

Cell membrane of ovum

Nucleus containing chromosomes

Acrosome containing enzymes

(1)

(2)

(3)

(4)

(5)

Fertilization
Figure 20.2

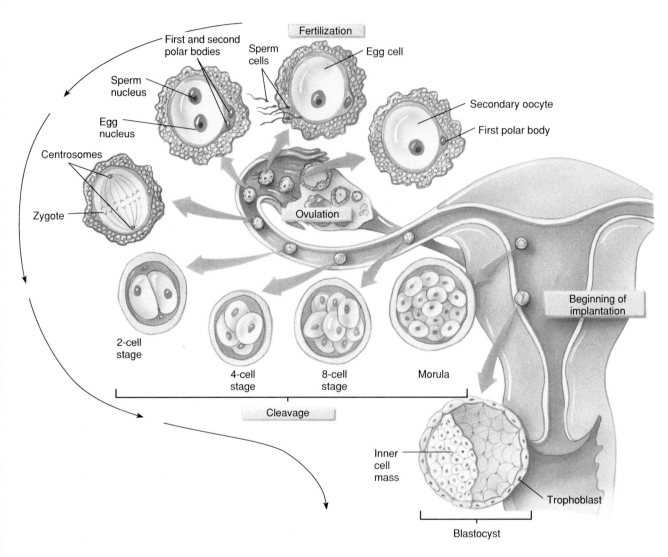

Early Human Development
Figure 20.3

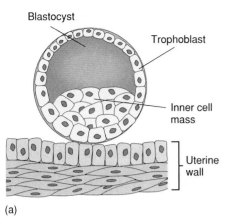

(a)

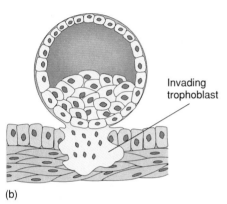

(b)

Blastocyst Implantation
Figure 20.5

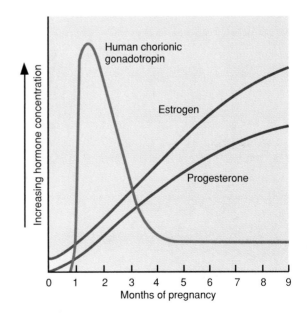

Hormone Concentration During Pregnancy
Figure 20.6

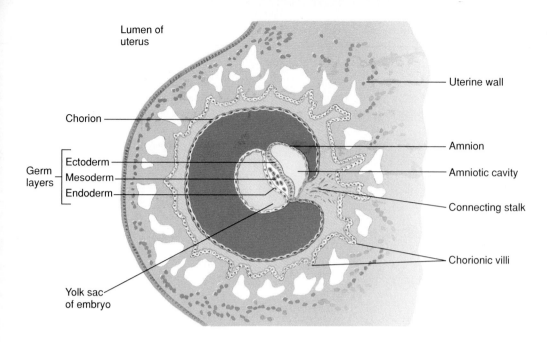

Primary Germ Layer Formation
Figure 20.7

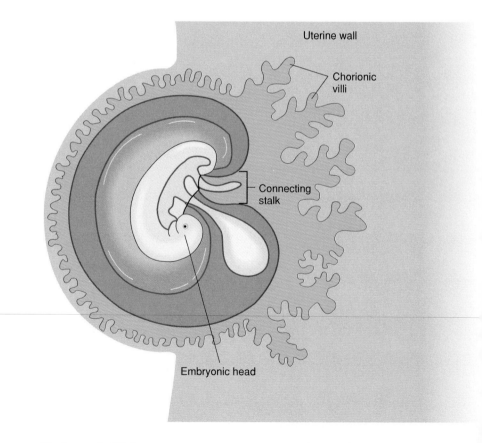

Embryonic Disk
Figure 20.8

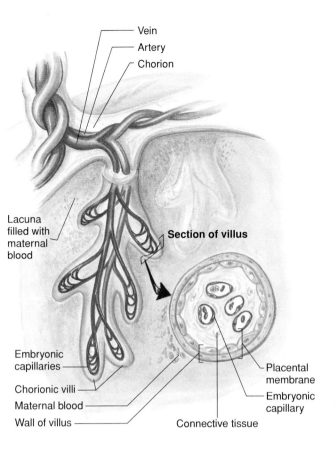

Vein
Artery
Chorion

Lacuna filled with maternal blood

Section of villus

Embryonic capillaries
Chorionic villi
Maternal blood
Wall of villus

Connective tissue

Placental membrane
Embryonic capillary

Chorionic Villi
Figure 20.11

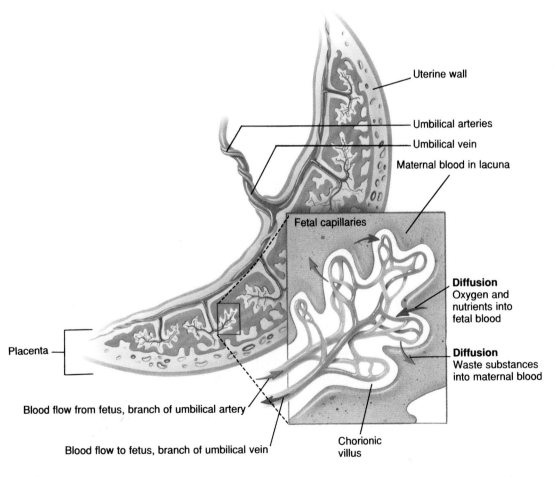

Uterine wall
Umbilical arteries
Umbilical vein
Maternal blood in lacuna

Fetal capillaries

Diffusion
Oxygen and nutrients into fetal blood

Diffusion
Waste substances into maternal blood

Placenta

Blood flow from fetus, branch of umbilical artery

Blood flow to fetus, branch of umbilical vein

Chorionic villus

The Placenta
Figure 20.12

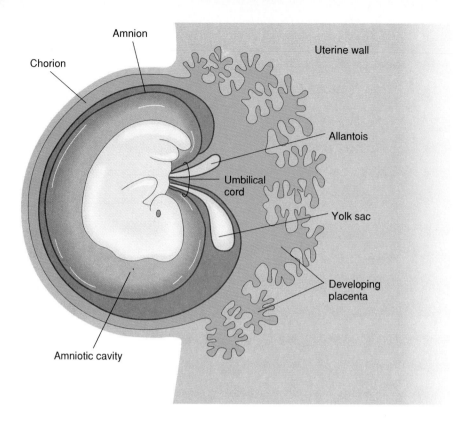

Umbilical Cord Formation
Figure 20.13

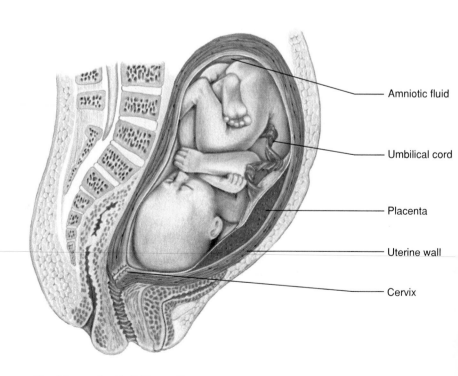

Position of a Full Term Fetus
Figure 20.15

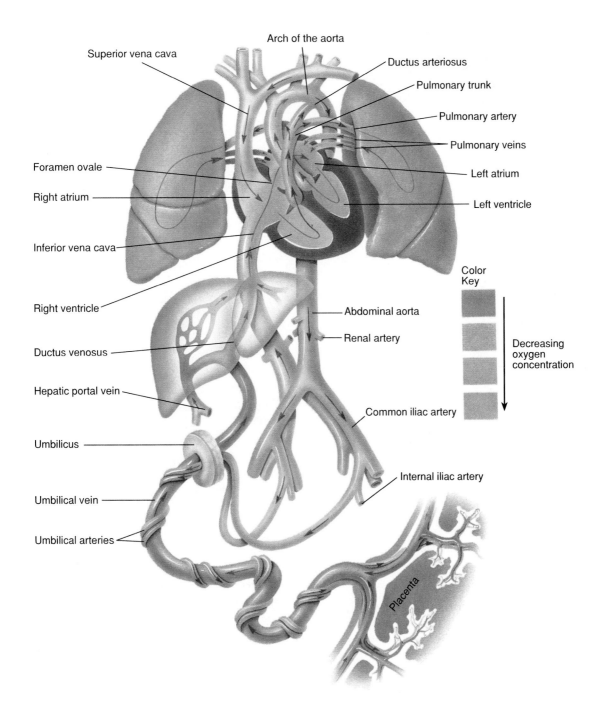

Superior vena cava

Arch of the aorta

Ductus arteriosus

Pulmonary trunk

Pulmonary artery

Pulmonary veins

Foramen ovale

Left atrium

Right atrium

Left ventricle

Inferior vena cava

Right ventricle

Abdominal aorta

Renal artery

Ductus venosus

Hepatic portal vein

Common iliac artery

Umbilicus

Internal iliac artery

Umbilical vein

Umbilical arteries

Placenta

Color Key

Decreasing oxygen concentration

Fetal Circulation
Figure 20.16

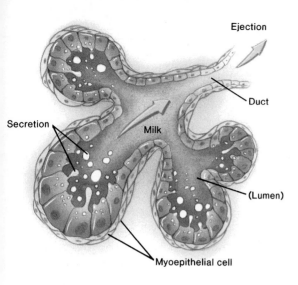

Milk Ejection
Figure 20.18

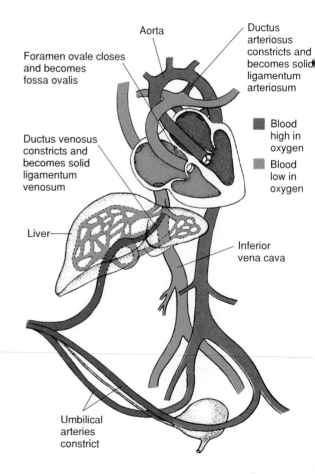

Changes in Newborn's Circulation
Figure 20.19

CREDITS

Line Art

Fig. 3.16 From Stuart Ira Fox, *Human Physiology,* 4th ed. Copyright © 1993 The McGraw-Hill Companies, Inc. All Rights Reserved. Reprinted by permission.

Fig. 6.1 From Kent M. Van De Graaff and Stuart Ira Fox, *Concepts of Human Anatomy and Physiology,* 4th edition. Copyright © 1995 The McGraw-Hill Companies, Inc. All Rights Reserved. Reprinted by permission.

Fig. 7.7 From Kent M. Van De Graaff, *Human Anatomy,* 3d ed. Copyright © 1992 The McGraw-Hill Companies, Inc. All Rights Reserved. Reprinted by permission.

Fig. 7.8 From Kent M. Van De Graaff, *Human Anatomy,* 3d ed. Copyright © 1992 The McGraw-Hill Companies, Inc. All Rights Reserved. Reprinted by permission.

Fig. 7.10 From Kent M. Van De Graaff, *Human Anatomy,* 3d ed. Copyright © 1992 The McGraw-Hill Companies, Inc. All Rights Reserved. Reprinted by permission.

Fig. 7.11 From Kent M. Van De Graaff, *Human Anatomy,* 3d ed. Copyright © 1992 The McGraw-Hill Companies, Inc. All Rights Reserved. Reprinted by permission.

Fig 7.12 From Kent M. Van De Graaff, *Human Anatomy,* 3d ed. Copyright © 1992 The McGraw-Hill Companies, Inc. All Rights Reserved. Reprinted by permission.

Fig. 7.13 From Kent M. Van De Graaff, *Human Anatomy,* 3d ed. Copyright © 1992 The McGraw-Hill Companies, Inc. All Rights Reserved. Reprinted by permission.

Fig. 8.14 From Kent M. Van De Graaff and Stuart Ira Fox, *Concepts of Human Anatomy and Physiology,* 4th ed. Copyright © 1995 The McGraw-Hill Companies, Inc. All Rights Reserved. Reprinted by permission.

Fig. 8.15 From Kent M. Van De Graaff and Stuart Ira Fox, *Concepts of Human Anatomy and Physiology,* 4th ed. Copyright © 1995 The McGraw-Hill Companies, Inc. All Rights Reserved. Reprinted by permission.

Fig 9.4 From Kent M. Van De Graaff, *Human Anatomy,* 4th ed. Copyright © 1995 The McGraw-Hill Companies, Inc. All Rights Reserved. Reprinted by permission.

Fig. 9.24 From Kent M. Van De Graaff, *Human Anatomy,* 4th ed. Copyright © 1995 The McGraw-Hill Companies, Inc. All Rights Reserved. Reprinted by permission.

Fig 12.18 From Ricki Lewis, *Human Genetics.* Copyright © 1994 The McGraw-Hill Companies, Inc. All Rights Reserved. Reprinted by permission.

Fig. 13.28 From Kent M. Van De Graaff and Stuart Ira Fox, *Concepts of Human Anatomy and Physiology,* 4th ed. Copyright © 1995 The McGraw-Hill Companies, Inc. All Rights Reserved. Reprinted by permission.

Fig. 14.2 From Kent M. Van De Graaff, *Human Anatomy,* 3d ed. Copyright © 1992 The McGraw-Hill Companies, Inc. All Rights Reserved. Reprinted by permission.

Fig. 15.3 From Kent M. Van De Graaff and Stuart Ira Fox, *Concepts of Human Anatomy and Physiology,* 4th edition. Copyright © 1995 The McGraw-Hill Companies, Inc. All Rights Reserved. Reprinted by permission.

Fig. 15.9 From Kent M. Van De Graaff, *Human Anatomy,* 4th ed. Copyright © 1995 The McGraw-Hill Companies, Inc. All Rights Reserved. Reprinted by permission.

Fig. 15.28 From Kent M. Van De Graaff and Stuart Ira Fox, *Concepts of Human Anatomy and Physiology,* 4th edition. Copyright © 1995 The McGraw-Hill Companies, Inc. All Rights Reserved. Reprinted by permission.

Fig. 17.7 From Kent M. Van De Graaff and Stuart Ira Fox, *Concepts of Human Anatomy and Physiology,* 4th edition. Copyright © 1995 The McGraw-Hill Companies, Inc. All Rights Reserved. Reprinted by permission.

Fig. 19.2 From Kent M. Van De Graaff, *Human Anatomy,* 2d ed. Copyright © 1988 The McGraw-Hill Companies, Inc. All Rights Reserved. Reprinted by permission.

Fig. 19.11 From Kent M. Van De Graaff and Stuart Ira Fox, *Concepts of Human Anatomy and Physiology,* 4th edition. Copyright © 1995 The McGraw-Hill Companies, Inc. All Rights Reserved. Reprinted by permission.

Fig. 19.14 From Kent M. Van De Graaff and Stuart Ira Fox, *Concepts of Human Anatomy and Physiology,* 4th edition. Copyright © 1995 The McGraw-Hill Companies, Inc. All Rights Reserved. Reprinted by permission.